TESLA

FOR BEGINNERS®

BY

ROBERT I. SUTHERLAND-COHEN

ILLUSTRATED BY

OWEN BROZMAN

FOREWORD BY

JANE ALCORN

FOR BEGINNERS®

For Beginners LLC
155 Main Street, Suite 211
Danbury, CT 06810 USA
www.forbeginnersbooks.com

Cataloging-in-Publication information is available from the Library of Congress.

ISBN # 978-1-939994-48-6 Trade

Manufactured in the United States of America

For Beginners® and Beginners Documentary Comic Books® are published by For Beginners LLC.

First Edition

10 9 8 7 6 5 4 3 2 1

To Patricia – The True Light of My Life

CONTENTS

FOREWORD

by Jane Alcorn

Celebrated inventor, lauded visionary, creative genius, and desired guest at social functions. Lonely man, forgotten person, rejected fringe scientist, ridiculed idealist, crazy bird-lover. Rediscovered scientist, inspiration to many, honored builder of the future, fascinating character. All of these labels apply to the same person at different times.

Nikola Tesla, recognized and honored for achievements early in his career, forgotten and ignored toward the end of his life, and noticed again in recent years, has become a hero to fans all over the world. Rediscovering the contributions to society made by this formerly unsung genius, people are now becoming aware of how his work underpins so much of the technology we take for granted today.

Tesla's work and ideas continue to inspire. As president of Tesla Science Center at Wardenclyffe—the site of his last laboratory on Long Island, New York—I have learned first-hand of the unique appeal of this Serbian-born innovator and eccentric. When we first set out to save Wardenclyffe, it was just to save a piece of history and start a small museum where children could learn about science. But the more we learned about the man whose vision brought us forward with his technological breakthroughs, the more interested we became in Tesla himself. To our delighted surprise, the rest of the world agreed that Tesla was important, too. Contributions came from across the globe. His name and renewed fame brought together over 33,000 people from 108 countries around the world who contributed to saving his cherished final worksite.

Tesla sought to make life better for everyone. His work was dedicated to answering the questions he had, and to building the technology we would need for the future. Concerned about our finite resources, about efficient use of energy, and about improved communications between people, he thought about and developed devices, technology, and systems to improve our lives.

From the alternating current system of electrical distribution upon which all of society depends, to alternators for cars, flow meters for aquatic vessels, robotics, remote control, radio, and so many other inventions and concepts, Tesla has led the way in technology. Over a hundred years ago he envisioned a hand-held device that could connect people all over the globe with pictures, voice, music, and information. Today we hold in our hands a smart phone that embodies that very idea. Free energy from the earth and wireless distribution of energy—other goals of his—are still to be realized, but scientists have begun to work on those possibilities.

Tesla is also a character in current television and film productions. From network channels to public television to cable stations, many programs have been devoted to Tesla's life and work. Feature films have been and continue to be written, developed, and made about him, or including him. Others are in the planning stages.

With a car company bearing his name, a science center being created in his former laboratory, new investigations being made into his science, and films being made about his life, Tesla is again the fascinating figure he once was.

A true futurist, Tesla realized that people wouldn't necessarily understand him in his own time. But it was for the future, now being realized, that he worked. In his own words,

> *Let the future tell the truth, and evaluate each one according to his work and accomplishments. The present is theirs; the future, for which I have really worked, is mine.*

With all the new interest in Nikola Tesla and his work, the demand for information is growing. There have been several scholarly works published in recent years that reflect this interest, but they tend to be aimed at those who want in-depth, heavily-footnoted, detailed information. This volume is written with students and general readers in mind. *Tesla For Beginners* provides an introduction to the man and his accomplishments. In this work of graphic nonfiction, students of any age can find solid information and background about a genius who gave so much to the world.

Jane Alcorn is President of the Tesla Science Center at Wardenclyffe, Shoreham, NY (www.teslasciencecenter.org)

Introduction

OVERVIEW OF A GENIUS

STOP. BEFORE YOU READ ANY FURTHER, take a few moments and note all the devices that are powered by electricity within 10 feet of your body's radius. You will be astonished by the number. Imagine how this number grows exponentially as you broaden the circle beyond the room, throughout the building, around the block, and out across the world. It is hard to imagine that one man was responsible for inventing the process, the machinery, and the distribution network that delivers alternating current (AC) to power all these devices.

Oh, by the way, are you reading this book under fluorescent lighting? Or perhaps you are listening to a radio at the same time. Maybe a television is on. And is that a cell phone ringing in your pocket? These are just some of the inventions or devices that employ ideas or circuitry invented by this same man, who filed nearly 300 patents in his lifetime. Many of these patents resulted in functioning inventions. Others were little more than wide-eyed dreams—or still await possible development.

The man behind alternating current and wireless technologies traveled from Serbia by steamship to arrive in the United States with only four cents in his pocket. It was in the early 1880s, at the tail end of the Industrial Revolution and the beginning of the Second Industrial Revolution, that America beckoned him.

Into this age of wild, individual invention strode an idealist with a head full of potential improvements that would radicalize the distribution of electrical current, boost the well-being of society, and contribute to world peace. He would tangle with some of the most brilliant minds in America's history, including Thomas Edison, George Westinghouse, J.P. Morgan, and Stanford White. His social circle was legion, among them Mark Twain, Robert Underwood Johnson and his wife Katharine, poets, musicians, and writers. Such notables flocked around him in hopes of profiting one way or another from his plans for global wireless communication systems.

He achieved financial wealth and worldwide recognition even though he outlined some disastrous proposals that never came to fruition. Ultimately he took a precipitous fall. He spent his last decades scraping for funding for celestial projects and living out his final days in penurious solitude with a pigeon.

This poet of invention, possessed of a prodigious memory and deep-seated phobias, has left behind a vast and intriguing legacy. He was a scientist, a physicist, a mathematician, an electrical engineer, a writer of verse, and an extensively published author.

So let's explore the enigmas, visions, dreams, and plausible realities of this towering but sometimes overshadowed figure of modern science and invention: NIKOLA TESLA.

Chapter 1
BEGINNINGS

And God said, Let there be light . . .
—Genesis 1:3

Precisely at the stroke of midnight, as the clock turned from July 9, 1856, to July 10, 1856, Nikola Tesla was born in the village of Smiljan, in the mountainous Austrian province of Lika. (Lika is a historical region of Croatia and is located today in the Republic of Croatia. At the time of Tesla's birth, however, it was part of the Austria-Hungarian Empire). As both lore and his own autobiographical writings tell us, Tesla came into being amid a raging lightning storm. Whether or not he was possessed with God's prophecy, it was indeed his destiny to light up the world. And while he is credited with achieving just such a mythical feat, he was first and foremost a man of nature, scientific inquiry, and prodigious invention. A view into his early years, inventive process, phobias, reasoning powers, reclusiveness, dreams, and survival tactics comes to us through his autobiography, *My Inventions* (1919).

Homeland

The world of Smiljan has a long, volatile history, caught up in what we now know as the warring Balkan states. Lika and other regions in Croatia were under Austria-Hungarian rule beginning in the mid-1800s, while neighboring Bosnia was under Turkish control. Both Croatia and Bosnia fought for years to gain independence from their ruling powers. Once they obtained their freedom—and right up to the present day—the diverse factions in Croatia, Serbia,

Bosnia, and Slovenia continued to fight among themselves and their neighboring states. Following World War II, Croatia became one of the six constituent republics of the Yugoslav socialist federation. Once again, Croatia fought for its freedom and engaged in bitter fighting with Serbia. The complex of shifting borders across this region of the world gave rise to the word *balkanization.*

> **BALKANIZATION**
>
> The division of an area, country, region, or group into smaller and often mutually quarrelsome units.

It was in this geopolitical context and amid these natural surroundings that Nikola Tesla spent his childhood. Smiljan lay across the western edge of the Austria-Hungarian Empire's military frontier. Many of its men were conscripted to go off and fight wars, while others chose a religious path. Meanwhile, the female residents back home eked out a living on hardscrabble farms, often plagued by famine. By the time of Tesla's birth, a decline in Turkey's control of the surrounding area gave rise to civilian administration.

Although Tesla was born in Croatia, his parents were of Serbian descent. Milutin, his father (1819–1879), was a minister who graduated at the top of his class. He had no desire to pursue a military career, as so many of the men in his family had done. Tesla's mother, Djuka (translated as Georgina) Mandic (1822–1892), was the daughter of a priest. Her family lineage included many who chose a career in the clergy. Shortly after their marriage in 1847, Milutin was transferred to a parish in Senj and later to one in Smiljan.

Milutin was a pastor, writer, and poet with an extensive library on wide-ranging subjects. He strongly desired independence from the Austrians and Turks, as well as everlasting peace. Milutin spoke many languages, was adept at mathematics, and trained his sons to perform calculations in their heads as well as feats of memory.

Although Djuka could not read, she committed to memory long passages from the Bible. She was a dutiful homemaker, who in her own right was an inventor of many household items. Working from sunup to late in the evening, she would run the farm and household. Her sewing abilities were renowned, aided by looms she devised herself. To help prepare meals, she invented churns and labor-saving kitchen devices for their small, isolated farm. Both parents shared a sense of the spiritual and viewed life beyond the rigors of farming. Their quest to improve the human condition was not lost on the young Tesla. The fourth of five children, he was closest to his mother and shared her work ethic, as well as her sense of invention for tools that would better the world.

Early Visions and Scientific Intuition

"Our first endeavors are purely instinctive," Tesla would write in his autobiography, "prompting of an imagination vivid and undisciplined. As we grow older, reason asserts itself and we become more and more systematic and designing. But those early impulses, though not immediately productive, are of the greatest moment and may shape our very destinies."

A number of events in Tesla's childhood life were to have a marked effect on his future as an inventor, dreamer, strategist, and communicator in different realms. Young Nikola spent much of this early life amid birds, chickens, geese, sheep, horses, and cats. It was an extensive immersion in the natural wonders around him. He went to great lengths to talk to the animals, especially his favorite cat, Mačak.

Stroking Mačak on a cold, dry winter night, Nikola observed sparks of light emanating from the animal's backside. His father

explained that this was static electricity, much like that seen during lightning storms. It was the same kind of electricity one was able to produce by rubbing one's feet on a rug, resulting in a spark when touching another person. Tesla was fascinated. He reasoned that the jolt of electricity was a kind of power or energy, and he wondered: How could he produce greater quantities of electricity? What would he have to do increase the output of these electrical forces? Then, how could he utilize electricity to power machinery? As he recalled at the age of 80, these were all grand ideas circulating in his head at an early age.

> *I cannot exaggerate the effect of this marvelous night on my childish imagination. Day after day I have asked myself "what is electricity?" and found no answer. I still ask the same question, unable to answer it. Some pseudo-scientist, of whom there are only too many, may tell you that he can, but do not believe him. If any of them know what it is, I would also know, and my chances are better than any of them, for my laboratory work and practical experience are more extensive, and my life covers three generations of scientific research.* ("A Story of Youth Told by Age")

In addition to communing with the farm animals, Nikola was close to his elder brother Dane (b. 1848) and formed more distant relationships with his sisters, Angelina (b. 1850), Milka (b. 1852), and Marica (b. 1858). Much of his playtime was spent intermingling with nature, observing cause and effect. The forces of nature—such as the power of flowing water and wind—were very real to him. In particular, the impact of water current on floating objects took shape in his mind. How would he transform these myriad thoughts into reality and useful purposes? These were his seeds of invention. From a young age, he began formulating ideas for translating natural forces into the transmission of usable energy.

By observing how a flowing stream could propel a toy boat, for example, he began to formulate variations of the physical event. For a budding inventor, this was the magic "what if?" moment. Nikola theorized that the water's current would have the same effect on a wheel or disk, which would rotate if mounted on an axle hovering perpendicular above the flowing stream. A portion of the disk just below the axle would be immersed in the water and would spin in direct relationship to the current pushing against it. With that theory in mind, Nikola constructed just such an apparatus and noted the resultant effects. How his observations would figure in his future would require the test of time. For the moment, these were early hints of his thought process for the invention of a smooth-disk turbine (PATENT 1,061,142 – FLUID PROPULSION, filed October 21, 1909, and PATENT 1,061,206 – TURBINE, filed October 21, 1909).

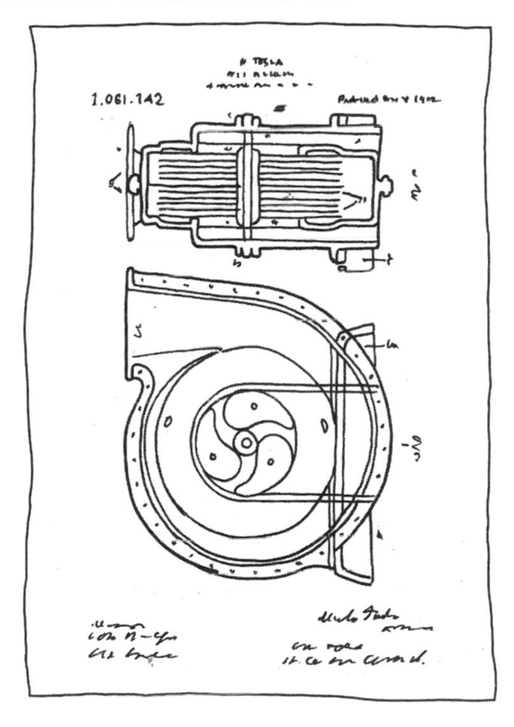

Air flight was another concept that loomed large in the boy's mind, so much so that he took to leaping off the barn roof with umbrella in hand to ease his descent. He spent a number of months in bed healing.

In addition to experimenting with physical phenomena, Tesla believed strongly that he possessed the ability to both communicate and predict events beyond the natural order of things. He was convinced that his sensory apparatus went beyond seeing and hearing the things right in front of him. He claimed that he could hear a small object dropped in another room or at a great distance, and that the sound of unseen objects at times became so deafening that it caused him intolerable headaches. On one occasion, he visualized the death of his cat Mačak. When later told the exact details of his pet's passing, the account confirmed precisely what he had seen in his mind.

Similarly, in thinking about a possible invention, Tesla would pose a series of questions to himself and work through the potential solutions. Ultimately, release would come in the form of what he regarded as complete clairvoyance for a practical invention. At other times, he imagined that he could reach right through the space in front of him and touch aspects of the image. If he saw something, he wondered, would it be possible to invent a device that could project the image as his eye saw it? Obsessively he grappled with these ideas. And though nothing came of many, they were the earmarks of the intense scientific inquiry that gripped Tesla from a very early age.

Another unique childhood invention came about through his observation of the forces of wind. Living on a farm, it would not be far-fetched to assume that he was able to observe the effects of a windmill in the threshing of grain. But, young Tesla wondered, what if there is not enough wind to turn the mill?

Melding what he learned from the water-driven spinning wheel, he devised a system of pulleys that attached a wheel to a model windmill. Then he tethered a collection of May flies (June bugs in North America) to the windmill's vanes. The frantic flapping of the bugs' wings set the wheel in motion. Nikola proudly showed off his invention to a boy in the neighborhood who proceeded to eat the bugs. The sight of him gobbling the bugs so disgusted Tesla that he vowed never to harm an animal again. And according to one biographer, he never repeated the experiment in the rest of his life.

Nikola eagerly moved on to other juvenile experiments. He would tinker endlessly with clocks, taking them apart but never succeeding in putting them back together. As military weapons caught the fancy of many a young boy, he constructed a pop-gun in which he utilized forced air-pressure to propel a wad of hemp. The result was a few shattered windows. In another flight of invention and imagination, he fashioned his own swords and waged war against an imaginary advancing army of cornstalks. Such adventures aroused the ire of his parents, but for Nikola they were the beginnings of his life-long search for what makes things work.

One particular image came to haunt him for the rest of his life. It occurred when he witnessed his brother Dane being thrown by their favorite horse. The resulting injuries led to his brother's death at the age of 12. The incident replayed itself in his mind, in stark detail, throughout Tesla's life. Dane's death was a cause of great anxiety and frequently distress. Tesla's fixations would lead to crippling headaches and mental breakdowns on numerous occasions. Often his memories would be accompanied by brilliant flashes of light inside his mind, as if war was being waged in his head.

Early Education and the Possibility of Advanced Power

Young Nikola Tesla had looked up to his brother. Dane's death in 1861 was a stunning blow not only to young Nikola; it proved devastating to their parents. No matter what Nikola did, he felt that he lost favor with his parents. His mother seemed to reject whatever he tried to do. His father Milutin had grown so despondent that he carried on conversations with himself, changing voices for the imagined participants in the conversation. Eventually he requested a transfer to another parish. When the request was granted, the family moved to the nearby town of Gospić. The move turned Nikola's world upside-down. It broke his heart to part with his pigeons, chickens, sheep, and magnificent flock of geese. He felt like a prisoner in his new house and grew increasing bashful around outsiders.

Various eccentricities and obsessive behavioral patterns also began to emerge in Tesla around this time. Some of these were manifest throughout his life. He reacted violently to the sight of women's earrings. Pearls gave him fits. He was fascinated by glittering crystals or objects with sharp edges and plane surfaces. He would not touch the hair of other people. The sight of a peach would make him feverish. A meal would be joyless if he didn't calculate the cubical contents of soup plates, coffee cups, and pieces of food. Counting steps was another obsession, and all repeated acts he performed had to be divisible by three. If not, he would repeat the actions until they were.

EIDETIC MEMORY

Today psychologists refer to the kind of imaging that Tesla experienced as *EIDETIC MEMORY*. This is the ability to picture a past occurrence in extraordinarily vivid detail, sometimes triggering strong new thoughts or feelings.

One incident somewhat brought Tesla out of his shell and endeared him to the community. With much pomp and circumstance, the townspeople gathered at the river to unveil their recently purchased fire pump. Upon completion of the speeches and ceremonies, the command was given to pump. Not a drop of water issued from the nozzle, and no one on the scene could solve the problem. Knowing a little about air pressure, Nikola dove into the river and instinctively uncrimped the collapsed suction hose. Water gushed from the nozzle, making Tesla a local hero.

Nikola gained much of his knowledge from reading the many books in his father's library. His mind roamed far and wide as he immersed himself in the literature, poetry, and stories of distant lands. Milutin wanted Nikola to follow a clerical career path and feared that the boy would ruin his eyes from so much reading. Much to his father's consternation, Tesla would fashion candles, seal the keyhole and cracks around his bedroom door, and read in secret long into the night. The lack of sleep did little to affect his daytime activities, and it was a work pattern he would adapt to throughout much of his life.

Eventually his early fascination with flowing water and wind currents gave rise to more grandiose ideas of water turbines. He found a number of turbine models around his school and enjoyed building and operating others. A picture and description of Niagara Falls imprinted in his imagination a giant wheel run by the giant waterfall. So powerful was the image in young Tesla's mind that he boasted he would bring his vision to reality one day.

Junior High School—Formulating Theories

At the age of ten, Nikola Tesla enrolled in the Real Gymnasium (Junior High School) in Gospić. There he displayed an astonishing skill in mathematics—he could visualize and calculate long strands of mathematical problems in his head. So great was this proficiency that one professor accused him of cheating when he was able to rattle off solutions to problems without a piece of paper to do the calculations.

As good as he was at mathematics, however, his drawing ability was below par. Tesla felt little inclination to make pictures or diagrams, as the effort interfered with his thought process. The inability to draw threatened to spoil his career.

Worse, his daydreaming and inability to distinguish between reality and imagination became troublesome. Seeking the will to separate the real from the imagined, he began to search within himself. At the age of twelve, he stumbled across a Serbian translation of the romantic novel *Abafi: Son of Aba,* by Hungarian author Miklós Jósika. Tesla would later say,

> *This work somehow awakened my dormant powers of will and I began to practice self-control. At first my resolutions faded like snow in April, but in a little while I conquered my weakness and felt a pleasure I never knew before—that of doing as I willed.* (My Inventions)

Such mastery of will became a great meditative tool for Tesla, enabling him to accomplish many brainy feats throughout his lifetime.

The scientific apparatus and various electrical and mechanical models that Nikola discovered in the school's physics lab were especially intriguing to him. They offered him the opportunity to experiment with motors, electricity, and water turbines. His efforts were guided by extensive readings on electricity.

Meanwhile, he continued to harbor his dream of flying. He reasoned that if he were able to apply steady air pressure to a cylinder within a vacuum, he could produce continuous forward motion. Then, a shaft inserted into the cylinder with a propeller on the other end would provide lift.

With the idea of a flying machine fixed in his mind, young Tesla was able to take imaginary journeys around the world. Looking back, his idea was a forerunner of what we now call a helicopter. Indeed it would lead directly to the future invention of the Tesla turbine. However, there was a flaw in the apparatus that did not dawn upon the inventor until many years later. As he wrote in *My Inventions,*

> *It took years before I understood that the atmospheric pressure acted at right angles to the surface of the cylinder and that the slight rotary effort I observed was due to a leak. Tho [sic] this knowledge came gradually it gave me a painful shock.* (My Inventions)

Tesla's derring-do and heroics continued during his Real Gymnasium years, further illustrating his deductive reasoning powers on air pressure and water currents. As he would later put it,

> *An inventor's endeavor is essentially lifesaving. Whether he harnesses forces, improves devices, or provides new comforts and conveniences, he is adding to the safety of our existence. He is also better qualified than the average individual to protect himself in peril, for he is observant and resourceful.* (My Inventions)

Those instincts came dramatically to the fore in two near-death drowning incidents. The first came about out during a swimming prank played on friends. Diving under a wooden raft, he expected to disappear under water and resurface on the opposite side. Underneath the floating structure, however, he became disorientated. He tried repeatedly to resurface but kept banging his head on the underside of the raft. Almost at the point of blacking out, he realized that there would be airspace between the deck of the raft and the supporting crossbeams just beneath it. Gasping for air, he made his way from one pocket of air to the next and—after his friends had lost all hope for his survival—eventually resurfaced on the other side. It was a frightening way to learn about surface tension and air pressure.

In another act of death-defying bravado a few years later, Tesla swam out to a nearby dam. Generally, the water rose within an inch or two of the wall of the dam. On this day, however, the water flowed fast and strong over the top of the dam. As Tesla approached the edge, the currents began to carry him over the top. At the last moment, he was able to grab hold of the wall and hang on with both hands. But as the water pressure against his body increased, his strength began to give out.

Thinking quickly, he remembered a diagram he had seen on one of the basic principles of hydraulics: the pressure of a fluid in motion is proportionate to the area exposed. With that in mind, Tesla turned onto his left side. This made him like a swimmer slicing through the water, rolling from side to side with each arm stroke. With less of his body opposed to the water, turning onto his side had the effect of lowering resistance and improving propulsion. Being left-handed, Tesla was able to drag himself to safety on the opposite bank.

Aside from torn skin along his left side, Nikola survived with only a fever. Lying in bed in the weeks that followed gave him an opportunity to think about ways to utilize the energy differences in air pressure to drive water turbines and motors. Already taking shape in his mind was a continuous-motion machine that would work by maintaining steady air-pressure in a vacuum and harnessing (like his earlier flying machine or a windmill) the rush of incoming air.

Upon recovering, Tesla plunged into his scientific studies—so much so, in fact, that he again became dangerously ill upon completion of courses at the Real Gymnasium. Once again, he used his convalescence to read constantly. Included in his readings were the novels of Mark Twain. Upon meeting Twain a quarter-century later, Tesla moved the author to tears when he related how the books had spurred his recovery. That encounter was the basis for a longstanding relationship between the two men.

High School

Nikola's extensive readings prepared him to enter the Higher Real Gymnasium (High School) in Karlovac (Carlstadt), Croatia, and further his science studies. In Karlovac, he lived with his aunt (father's sister) and her husband, a former military officer who had participated in a number of battles. Life with them was sheer grief for young Tesla. They fed him sparingly, and he contracted malaria soon after arriving in this low, marshy area. He remained on a steady regiment of quinine during his entire stay.

To escape the misery, Tesla once again plunged into his studies. Under the influence of his physics professor, he became deeply absorbed in the study of electricity and mathematics. The professor demonstrated principles with apparatus of his own design, including a device in the shape of a freely rotating bulb with tinfoil coatings that would spin rapidly when connected to a static generating machine. Tesla was fascinated. The phenomena played itself over and over in his mind, as he longed to know more about these mysterious forces.

Tesla worked at his studies with reckless abandon. From an early age, he was conversant in a number of languages. By the time he graduated, he was fluent in Serbian, English, Czech, German, French, Hungarian, Italian, and Latin. He compressed his course work from the normal four years to three years.

Tesla wanted to return home to impress upon his father how engaged he was with the study of physics and electrical experimenting. This was the career path he had chosen. Hanging over the young man's head, however, was his father's insistence that he enter the priesthood. And so Milutin was not ready to welcome Tesla home, urging him instead to go on a long hunting trip and re-think his career choice. His ulterior motive was to shield Nikola from a cholera epidemic raging in Gospić.

CHOLERA

A bacterial disease usually spread through contaminated water. Cholera causes severe diarrhea and dehydration. If left untreated, it can be fatal in a matter of hours, even in previously healthy people.

Unaware of the epidemic, Tesla disobeyed his father and returned home. His years of malnourishment, a weakened immune system, and his previous bout with malaria made him an easy target for cholera. His illness was so severe that within nine months he was knocking at death's door. With the end nearing, Tesla said to his father,

"Perhaps I may get well if you will let me study engineering."

"You will go to the best technical institution in the world," Milutin replied.

As Tesla slowly recovered, his thoughts turned increasingly to engineering college. He and the family, however, overlooked the mandatory three-year military conscription looming before him. Milutin sent him off to the mountains and, in the intervening year, apparently pulled strings with relatives in the army to release the young Tesla from his obligations.

Trekking through the mountains, meanwhile, Nikola had visions for two of his most outlandish inventions to that time. Both tested the limits of his mathematical knowledge. First, he imagined a tube that would run under the ocean between Europe and the United States. The tube would be able to channel spherical mail containers prodded through by water pressure. (The flaw in his early thinking was that he did not account for the drag of water on the sides of the tube. Resistance was an error Tesla was able to factor into his later invention of a steam turbine.)

The second idea demonstrated similarly imaginative thinking and even more impractical execution. If he were able to build a ring around the earth at the Equator, Tesla hypothesized, anyone could hop on or off the ring to visit places anywhere around the spinning globe during a 24-hour span.

Once he had regained his health and his father was able to secure a deferment from military service, Tesla was finally positioned to embark upon his pursuit of an electrical engineering degree. His boyhood inventions, his early schooling, his ability to control his imagination, his sense of adventure, his maternal influence, and his father's religious precepts all prepared Nikola Tesla—and fired his resolve—to become an inventor.

Chapter 2

HIGHER EDUCATION—JOURNEY OF DISCOVERY

Of all things, I liked books best.
—Nikola Tesla

IN 1875, TO APPEASE HIS FATHER, Tesla entered the Joanneum Polytechnic School in Graz, Austria, under the pretext of becoming a professor of mathematics and physics. Living and studying more than 227 miles north of his hometown of Gospić, Tesla finally had the independence to pursue his training in electrical engineering, the life of an inventor, and all the courses that piqued his interest. In addition to the math and technology requirements, he took courses in such wide-ranging fields as analytical chemistry, mineralogy, machinery construction, botany, wave theory, optics, French, and English. His studies intensified as his interests broadened. He gained proficiency in nine languages and memorized long passages from Descartes, Goethe, Spencer, and Shakespeare. He typically worked seven days a week, from 3:00 A.M. to 11:00 P.M., getting by on four hours of sleep.

Among his favorite teachers at the Polytechnic were Dr. Moriz Allé, who taught integral calculus and specialized in differential equations, and Professor Jacob Poeschl, who held the chair of experimental and theoretical physics. Tesla would describe Poeschl as,

> *the most brilliant lecturer to whom I ever listened. He took a special interest in my progress and would frequently remain for an hour or two in the lecture room, giving me problems to solve, in which I delighted. To him I explained a flying machine I had conceived, not an illusionary invention, but one based on sound, scientific principles, which has become realizable thru my turbine.... Prof. Poeschl was a methodical and thoroughly grounded German. He had enormous feet and hands like the paws of a bear, but all of his experiments were skillfully performed with clock-like precision and without a miss.* (My Inventions)

Learning About Electricity

Professor Poeschl's teachings, in all likelihood, included a thorough progression from the earliest discoveries in electricity to the most current knowledge of the day. A review of some of these advances is intrinsic to an understanding of the principles leading up to Tesla's own theories and inventions. It is also important to reemphasize the key question that neither Tesla nor scientists before or after him have been able to answer: "What is electricity?" Indeed all we really know are the results of what electricity accomplishes.

Many centuries before Tesla stroked his cat Mačak to produce sparks and a hair-raising effect, the Ancient Greeks noticed that brushing amber with wool attracts very light materials such as straw. Much as Tesla's father Milutin described the effect to young Nikola, this was another example of static electricity.

For years, budding electricians would collect mysterious electrical "fluid" (for lack of a better term) in Leyden jars. Once the jar was filled to its electrical capacity, the experimenter could withdraw a charge and receive an electrical shock. Their hair could literally stand on end. This demonstration might have continued to be seen as a parlor trick were it not for Benjamin Franklin's kite-flying experiment in 1752, which proved that lightning was electrical in nature. Fortunately, Franklin was not struck by lightning that day and lived to write profusely on his electrical experiments.

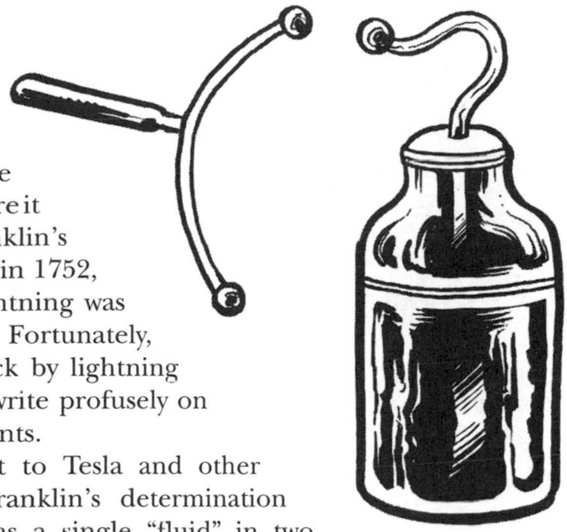

Especially important to Tesla and other experimenters was Franklin's determination that electricity flows as a single "fluid" in two different charged states, positive and negative. Franklin assumed that electricity flows from positive to negative, but that idea was disproved about 150 years later with the discovery of the electron. The negatively charged electron, it was shown, was the basis for the direction of current flow.

Franklin's invention of the lightning rod demonstrated that, if a metal wire is attached to the rod and strung down the side of a building and into the ground, the electricity generated by the lightning will run down the wire and into the earth, preventing damage to the building. The very notion of a lightning rod, the thousands of electrical discharges that take place during a thunderstorm, the erection of towering structures skyward, and the transmission of electrical energy all resonated deeply with Nikola Tesla. He would go on to conduct related experiments at Colorado Springs and Wardenclyffe at the end of the 19th century and into the 20th century, respectively.

The Physics of Electricity

Benjamin Franklin's electrical experiments pointed to the existence of electricity, but they did not establish what it is. Following Franklin's findings, there was increased activity across Europe to codify what electricity does. Reviewing the most important discoveries is vital to understanding how Tesla arrived at his own theories for electromagnetic induction to power machinery. Each successive discovery or new theory regarding the properties of electricity and magnetism resulted in a series of formulas. These formulas utilized an international system of units to represent certain base units (SI).

INTERNATIONAL SYSTEM OF UNITS

An internationally accepted coherent system of physical units, derived from the MKSA (meter-kilogram-second-ampere system). Scientists use the meter, kilogram, second, ampere, kelvin, mole, and candela as the basic units (SI units) respectively for the fundamental quantities of length, mass, time, electric current, temperature, amount of substance and luminous intensity.

Building on Franklin's description of the flow of electricity, the pioneering French physicist Charles-Augustin de Coulomb sought to provide a measurement of this flow. Knowing that certain materials, such as glass, rubber, and plastic, are electric **insulators** (charges can be rubbed on or off their surface and tend to stick there) and that other materials, such as copper, silver, and aluminum, are electric **conductors** (charges of electricity flow through them), Coulomb in 1785 defined a measurement for the number of charge carriers passed any given point in one second.

COULOMB

A simple household electrical circuit, such as a small "night light," carries a current of roughly 600 quadrillion (6×10^{17}) charge carriers per second! Engineers and scientists rarely write or speak about electrical current directly in terms of charge carriers per second. Instead, they use units of coulombs per second. A *COULOMB* (symbolized C) represents approximately 6.24×10^{18} unit negative or positive charges.

For Tesla or any other mathematician, 10^{17} condensed 10 to the 17th power and shortened a long string of numbers. Written out in long form, it would be 10 x 10 x 10 x 10 x 10 x10 x 10 x 10 x 10 x 10 x 10 x 10 x 10 x 10 x 10 x 10 x 10—i.e., 10 multiplied times itself 17 times successively. Given Tesla's prodigious memory and ability to do mathematical calculations in his head, we can assume that he was able to remember tables for powers

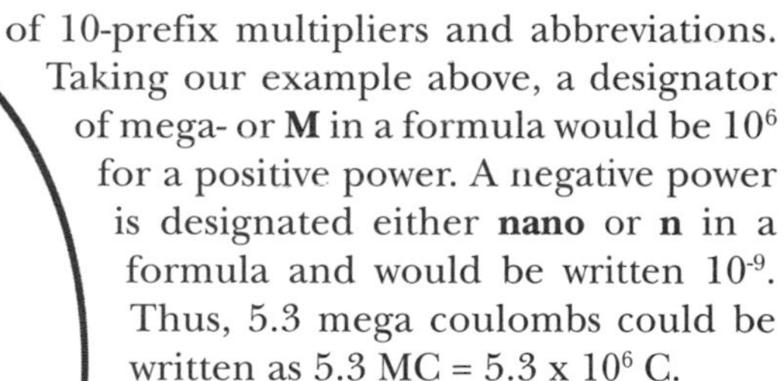

of 10-prefix multipliers and abbreviations. Taking our example above, a designator of mega- or **M** in a formula would be 10^6 for a positive power. A negative power is designated either **nano** or **n** in a formula and would be written 10^{-9}. Thus, 5.3 mega coulombs could be written as 5.3 MC = 5.3 x 10^6 C.

After André-Marie Ampere proved (1820) that wires could behave like magnets when a current passes through them, an *ampere* or *amp* (symbolized **A**) became the standard unit of measurement for electric current throughout the world. In SI base units, a current of 1 coulomb equals 1 ampere per second. Mathematically this would be expressed as:

$$1\ C = 1\ A/s$$

The early physics of batteries would be important to Tesla's understanding of how to actually store and produce electrical currents. In 1800, Italian Count Alessandro Volta built a pile of alternating copper and zinc metal disks, separated by paper pads soaked in salty water. He then attached separate copper wires to the top and the bottom of the pile. When he closed the circuit, a continuous electric current flowed through the pile. Known as the "Voltaic Pile," Volta's invention was the first battery. In his honor, the designation and measurement of electrical potential is called the *volt* (symbolized **V**). Today we see batteries of all strengths and purposes designated by their voltage (or number of volts), such as a 1.5-V penlight battery or a 13.5-V automotive battery.

In 1820, Hans Christian Oersted observed that when an electrical current passing through a voltaic pile is either stopped or started near the magnetic needle of a compass, the needle will be deflected. Here was proof positive of a relationship between electricity and magnetism. It is worth noting that, at the time of Oersted's observations, electricity began to be used for public lighting in only a limited way. The idea of powering motors or anything else was far in the future.

Building on Coulomb's demonstrations of electric charges repelling one another, as well as Oersted's and Ampere's discoveries that an electric current produces a magnetic field, the British chemist and electrical scientist Michael Faraday was able to invent the first primitive electrical motor (1831). Just as the lexicon of SI units grew with each new electrical attribution, so Faraday's work resulted in the adoption of the *farad* (symbolized **F**) as the base unit for electrical capacitance. Capacitance defines how much of an electrical field can be stored in an electrical component. The formula for one farad is:

$$1 F = 1 C / V$$

With the idea that a component can store electrical energy in the form of a magnetic field, Faraday reasoned that the component can also produce electrical energy. This was the landmark principle of electromagnetic induction. On the basis of that principle, it would be possible for a dynamo, or generator, to produce electricity by mechanical means.

INDUCTION

If an electrical current is passed through a coil of wire, the coil will produce a magnetic field around it. Similarly, if a magnetic field is introduced to the coil of wire, the coil will produce an electrical current. In these situations, we can say the coil has been induced, or is an inductor.

Little did Tesla know while learning the system of SI units that his own scientific work would someday lead to a standard unit for magnetic flux density: the *tesla* (symbolized **T**). In the meantime, Faraday's electromagnetic induction theory had to figure strongly in Tesla's mind as he was finishing his first year of college. It was a year of sleepless nights, spent studying, driving himself to achieve the highest grades, and gambling as a way to fit in with other students who considered him an oddball.

Nikola returned home for the summer, once again rundown. He wanted to impress his father with his scholarly achievements, but Milutin was at odds with him for his bad health and gambling. This type of behavior was completely against his father's religious principles.

Second-Year Studies—Thoughts of Alternating Current

Tesla's return to the Polytechnic Institute that fall freed him from his father's incessant criticism, as he immersed himself once again in Professor Poeschl's physics lectures. An especially exciting moment came with the arrival of a direct-current Gramme dynamo (an electrical generator that produces direct current, or DC, with the use of a commutator) from Paris. The device had the horseshoe shape of a laminated field magnet and a wire-wound armature with a commutator.

COMMUTATOR

A commutator is a split-ring device that reverses the current in the armature coil. By reversing the current at the point of the commutator that was attached by *COPPER BRUSHES* to the coil, the torque, or force acting on the DC motor, would remain constant.

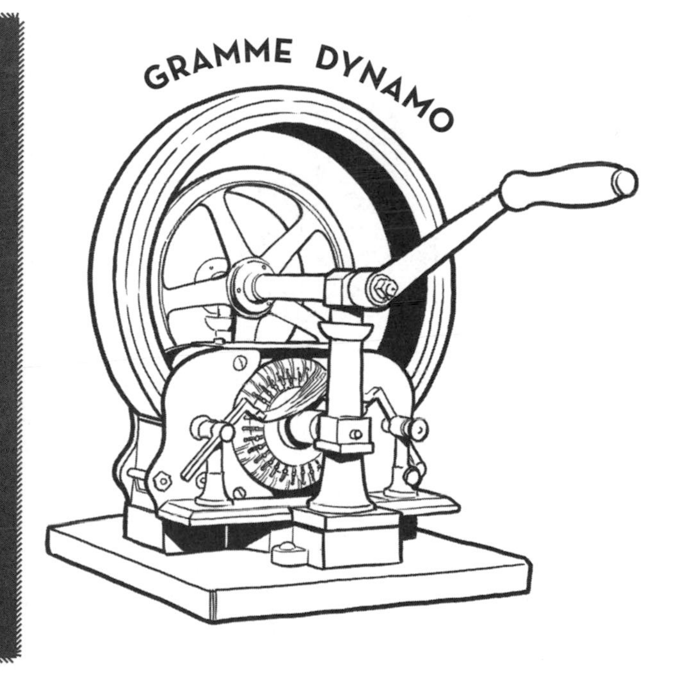

Intriguing as the dynamo was to Tesla, he had doubts that it was efficiently constructed. He was quick to point out its shortcomings.

The dynamo was connected up and various effects of the currents were shown. While Prof. Poeschl was making demonstrations, running the machine as a motor, the brushes gave trouble, sparking badly, and I observed that it might be possible to operate a motor without these appliances. But he declared it could not be done and did the honor of delivering a lecture on the subject, at the conclusion of which he remarked: "Mr. Tesla may accomplish great things, but he certainly never will do this. It would be equivalent to converting a steadily pulling force, like that of gravity, into a rotary effort. It is a perpetual motion scheme, an impossible idea." We have, undoubtedly, certain finer fibers that enable us to perceive truths when logical deduction, or any other willfull [sic] effort of the brain, is futile. For a time I wavered, impress [sic] by the professor's authority, but sooner became convinced I was right and undertook the task with all the fire and boundless confidence of youth. (My Inventions)

Professor Poeschl's embarrassing rebuke stung Tesla even more than his own father's admonishments. For the next few years, he pondered ways to prove Professor Poeschl wrong. There had to be a way to deliver alternating current without the use of a commutator and sparking brushes. To Tesla's way of thinking, a motor was only one part of a system. So he broadened his perspective. He eventually visualized a total system that included not only the motor, but a generator that would deliver the electricity to power the motor. The invention process would frustrate him for years before such a system would come to be realized.

College Dropout

Perhaps out of boredom, perhaps because he felt his professors were insensitive to his ideas about alternating current, Tesla adopted many of the dissolute ways of fellow students. He took to smoking, gambling, playing billiards, and drinking excessive amounts of coffee to keep pace with the group. His studies fell off, and he had to drop out of school during what would have been his third year in college.

Above all, Nikola did not want to face the wrath of his father. He was adrift. After trekking south to Maribor, near Graz, and taking a short-lived engineering job, he practically disappeared. Eventually Milutin was able to get in touch with him and tried to persuade him to return to the Polytechnic Institute. Failing this, his father urged Tesla to continue his studies at a college in Prague, the Charles-Ferdinand University. Instead, Tesla gambled away his father's money. As Tesla later recounted, his mother was more forgiving.

> *She understood the character of men and knew that one's salvation could only be brought about thru his own efforts. One afternoon, I remember, when I had lost all my money and was craving for a game, she came to me with a roll of bills and said, "Go and enjoy yourself. The sooner you lose all we possess the better it will be. I know that you will get over it."* (My Inventions)

The money she gave him quickly disappeared at the gambling tables. Dissolute, his studies aborted, Tesla soon came to recognize that his life would have to change. Here was another opportunity for him to apply free will, the lesson he had taught himself at age 12. As hard as it was for him, Tesla returned to the family home in Gospić and attempted to resolve the differences with his father by attending Milutin's church. As he continued to ponder his future, Nikola also made an effort to fit into the local society.

At 6 feet, 2 inches tall, he was a striking figure. He had a handsome face and was well-mannered, though perhaps a bit shy. He did not go unnoticed. In Gospić he attracted the attention of a young woman named Anna. Tesla espoused love, but thoughts of motors and alternating current were never far from his mind. With Milutin's death in April 1879, it was time to move on and honor his father's wishes that he complete a college education. Any potential love interest faded into the background.

Relatives contributed money to further Tesla's education in Prague, but their support was simply not enough to sustain full-time study. He continued to take courses but never received a degree, remaining a self-taught engineer with extensive language skills.

Budapest—Proving the Possibility of Alternating Current

Believing he could find employment at Alexander Graham Bell's telephone exchange in Budapest, Tesla set out for the Hungarian capital in early 1881. Unfortunately, plans for the Budapest exchange had not yet come to fruition.

In desperate need of money, Tesla sought help from relatives, who were able to secure him a draftsman's job in the Hungarian Central Telegraph Office. But the role was not well-suited to the 24-year-old Serb, who was terrible at drafting. It drove him to a complete nervous breakdown; he believed he was going to die. As he later explained the experience,

> *the period of that illness surpasses all belief. My sight and hearing were always extraordinary. I could clearly discern objects in the distance when others saw no trace of them. Several times in my boyhood I saved the houses of our neighbors from fire by hearing the faint crackling sounds which did not disturb their sleep, and calling for help.* (My Inventions)

It was an excruciating time for him. Any light or sound seemed to pierce his body or pound thunderously in his head. His heart would beat rapidly, his entire body would convulse.

To aid in his recovery, a former classmate and engineer named Anthony Szigeti would accompany Tesla on long walks. Szigeti became an instrumental sounding board for Tesla's theories. Their walks and talks in the city's parks slowly brought Tesla back to health, restoring his will to live. His energy renewed, Tesla resumed his mental quest to prove that an alternating current system without commutators or brushes was a viable alternative to motors driven by direct current. Solely in his mind, without a single piece of apparatus, he conducted experiment after experiment, testing each part down to the millimeter.

Given to hyperbole and a prolific writer throughout his life, Tesla recounts his breakthrough discovery in dramatic terms in his autobiography. He arrived at the revelation during an afternoon stroll through a Budapest park with his friend Szigeti. As Tesla recited lines from Goethe's *Faust* from memory, he was struck by an inspirational passage.

As I uttered these inspiring words the idea came like a flash of lightning and in an instant the truth was revealed. I drew with a stick on the sand the diagrams shown six years later in my address before the American Institute of Electrical Engineers, and my companion understood them perfectly. The images I saw were wonderfully sharp and clear and had the solidity of metal and stone, so much so that I told him: "See my motor here; watch me reverse it." I cannot begin to describe my emotions.... A thousand secrets of nature which I might have stumbled upon accidentally I would have given for that one which I had wrestled from her against all odds and at the peril of my existence. (My Inventions)

What Tesla diagrammed in sand that day would eventually change the course of history. He could now proudly proclaim that he was an inventor. "This was the one thing I wanted to be," he would write. "Archimedes was my ideal. I admired the works of artists, but to my mind, they were only shadows and semblances. The inventor,

I thought, gives to the world creations which are palpable, which live and work." Although Tesla's solution certainly disproved Professor Poeschl's admonition, it was still only preliminary and hypothetical. The world would have to wait for conclusive, refined drawings, which accompanied his address to the American Institute of Electrical Engineers in New York, in May 1888.

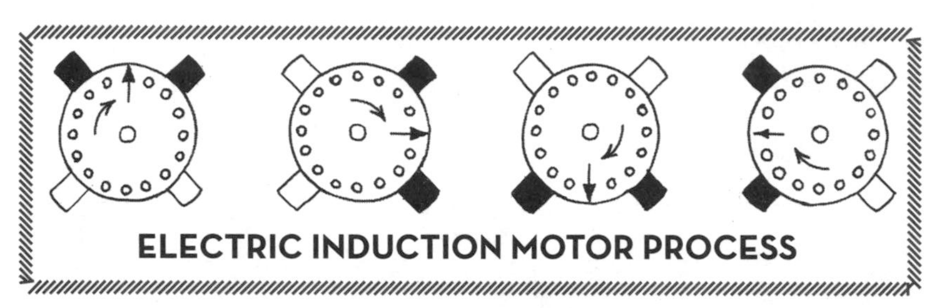

Essentially a motor works through the interaction of two sets of magnets, one stationary (the stator) and one able to move freely (the rotor). Both the stator and rotor are essentially coils of wire around metal cores. Rather than rely on permanent magnets that weaken over time, a motor utilizes electromagnets. Because electromagnets have no reason to set the rotor spinning, however, the polarity of the electromagnets has to keep switching in order to keep the motor turning. In early motors, this was accomplished by introducing a split-ring commutator to supply current to the rotor. Tesla's breakthrough was to remove the commutator and introduce only alternating current to the stator. The result was the creation of a magnetic field that opposed the magnetic field on the stator, thereby turning the rotor. Tesla's genius was discovering the perpetual back-and-forth motion that led to the creation of his induction motor.

Paris—Working for the Edison Organization

While buoyed by his breakthrough discovery, Tesla faced the reality of needing an income. With plans for the Budapest telephone exchange now in place, he was able to join the company in the latter part of 1881. He was a standout in his work, making all sorts of calculations and improvements for the various installations, including superior amplification systems. He rose rapidly in stature and was promoted by the owner of the central exchange, Ferenc Puskas, to the position of chief electrician. Tesla's contributions helped make the exchange a viable business, and eventually Puskas was able to sell it for a profit. Puskas in turn recommended Tesla and Szigeti to his brother, Tivador Puskas, who was introducing incandescent lighting systems for Continental Edison in Paris.

Here was just the opportunity Tesla had been waiting for. It would be a chance to meet the Edison people and introduce them to the rotating magnetic field. He accepted the position immediately and went to work for Continental Edison in Paris in 1882.

Living in Paris, Tesla stuck to a rigid daily schedule, rising before sunrise every morning, swimming in the Seine regardless of weather, and walking an hour to the company's factory in the suburb of Ivry. But there were many distractions in the beguiling City of Light. In the evenings he would play billiards with company colleagues, often discussing his electrical theories. He dined in some of the city's finest restaurants and soaked up Parisian culture in all its finery. Before the end of every month, he had spent most of his money. When Mr. Puskas asked him how he was getting along, Tesla replied, "The last twenty-nine days of the month are the toughest!"

Never far from his mind was a steady stream of mental calculations on how his discovery of the rotating magnetic field could solve many of the problems associated with the Edison system. The equipment was built or repaired largely on the basis of methods handed down by one employee or another, sometimes through much trial and error. The workers were not trained in physics or engineering; nor did they possess Tesla's knowledge of mathematics. When he would try to explain a problem from his perspective, no one was interested. This was a big disappointment for Tesla, as the company was more interested in the profits to be derived from spreading incandescent lighting than in electrical power motors.

Nevertheless, Tesla continued to observe how the various dynamos and motors were physically assembled and functioned at the plant. As was often the case, he wrestled with the theoretical configurations but did not physically assemble the equipment. With his vast mathematical and engineering knowledge, he was able to extrapolate formulas that would vastly improve the construction of the machines.

All his talk of inventions at Continental Edison was wasted until it caught the ear of a Mr. D. Cunningham, foreman of the Mechanical Department, who offered to form a stock company. "The proposal seemed to me comical in the extreme," Tesla later wrote. "I did not have the faintest conception of what that meant except that it was an American way of doing things."

Indeed nothing came of the proposal. Tesla thereafter shuttled back and forth between constituent Edison companies in France and Germany, repairing equipment at the various power plants. During one of his return trips to Paris, a company administrator named Mr. Rau offered Tesla the opportunity to improve the method for regulating current on its dynamo electric machines. He reveled in the task of developing automatic regulators for the dynamos (PATENT 336,961 – REGULATOR FOR DYNAMO-ELECTRIC MACHINES, filed May 18, 1885, following Tesla's arrival in the United States).

Although his invention did not dispense with commutators and brushes, Tesla's solution was so successful that the Edison Company entrusted him to solve a major problem with a lighting installation at the railroad station in Strasbourg, Germany, in 1883. As he explained the situation, "The wiring was defective and on the occasion of the opening ceremonies a large part of a wall was blown out thru a short-circuit right in the presence of Old Emperor William I. The German Government refused to take the plant and the French Company was facing a serious loss."

Tesla labored day and night to fix the system. His technical knowledge, work experience, and ability to speak German made him the right man for the job. Yet fixing the problem at the train station was hardly the greatest success in his own mind during his stay in Strasbourg. During his spare time there, he constructed his first alternating-current motor out of materials he brought from Paris, along with a disk of iron and bearings made for him in a local mechanical shop. "It was a crude apparatus," he remarked. "The consummation of the experiment was, however, delayed until the summer of that year when I finally had the satisfaction of seeing rotation effected by alternating currents of different phase, and without sliding contacts or commutators, all as I had conceived a year before."

PHASE

A phase is a property of waves. In its purest form, AC is represented by a sine wave. The sine wave has peaks and valleys, with a zero-crossing between them. A phase of the AC sine wave is the distance, or period, between the first zero-crossing and the point in space where the wave originated.

Eventually Tesla would formulate theories for polyphase (multiphase) systems that would generate power by utilizing alternating voltages of the same frequency, in which the phases are cyclically displaced by fractions of a period.

CURVES SHOWING CYCLE OF OPERATIONS

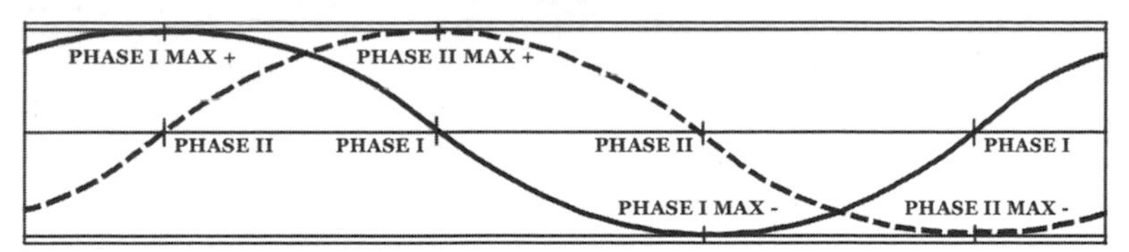

Tesla met a number of influential people while in Strasbourg and solicited them for capital for his new invention. Disappointed, he returned to Paris in early 1884 in the hope of at least some reward from Continental Edison for his successes in Strasbourg. Once again he faced disillusionment, as he was shuttled from one manager to another without receiving the payment he thought had been promised. His requests to demonstrate his alternating-current system and new motor met with disapproval as well, so the independent-minded young inventor quit the French Edison company.

Fortunately for him, a company administrator named Charles Batchelor—also a personal friend of Thomas A. Edison—urged Tesla to go to the United States and work for the firm on the design and manufacturing of dynamos and motors. Batchelor gave Tesla a personal letter of recommendation to Mr. Edison himself.

Off to America

Following Batchelor's suggestion, Nikola Tesla set out on his journey in spring 1884. He sold all of his possessions except for a few essential personal effects and used the money to buy a train ticket to the docks, where he was booked for transatlantic passage on the S.S. *Saturnia.*

He arrived at the station just as the train was pulling out, only to discover that his tickets and money were missing. Racing alongside the train, he jumped aboard at the last possible moment. With the few coins remaining in his pocket, he managed to pay the conductor for his ticket.

At the docks, Tesla explained his plight to skeptical steamship officials and was permitted to embark only when no one else showed up to claim his reservation. With the remnants of his belongings, a few poems and articles he had written, and a package of calculations for a problem related to flying machines, Tesla began his journey. He spent most of the trip in solitude at the stern of the ship, mind abuzz with thoughts of alternating current and his induction motor. He was off to the land of promise.

Chapter 3

THE PHOENIX RISES

The scientific man does not aim at an immediate result. He does not expect that his advanced ideas will be readily taken up. His work is like that of the planter—for the future. His duty is to lay the foundation for those who are to come, and point the way.

—Nikola Tesla

TESLA ARRIVED IN NEW YORK IN JUNE 1884 with four cents in his pocket, and the America he encountered seemed rough, crude, and a century behind Europe. The police were curt when he asked for directions. Cab fare was more than four cents, so he started walking. When he encountered a man in front of his shop kicking a machine, Tesla volunteered to fix it and was handsomely rewarded with a twenty-dollar bill.

Working for the Wizard

Nikola Tesla's working life in America was to begin the next day. With his letter of recommendation in hand, he was off to meet the great Thomas A. Edison. Batchelor's introduction read as follows: "I know two great men and you are one them; the other is this young man [Tesla]."

Tesla was thrilled at first to be working for the famous American inventor. But the 28-year-old Tesla and 37-year-old Wizard of Menlo Park were a contrast in styles. Tesla brought with him courtly Old World manners and refined European dress. He was fluent in English, well-read, full of hygienic phobias, and anxious to learn as much as he could about American customs. Edison, by contrast, was often gruff, dressed in plain home-made clothes, and spent the workday in a workman's lab coat. He was a boastful propagandist and would let little stand in his way.

A brilliant inventor and innovator, Edison was an aggressive businessman who built profitable commercial enterprises through guile, cunning, and a focus on profits. He supplied direct-current electricity for lighting some of the most opulent mansions of Lower Manhattan, as well as certain factories, theaters, and public venues around the city. His empire consisted of the Edison Machine

Works on Goerck Street downtown and the Edison Electric Light Company on Fifth Avenue. His generating station at 255-57 Pearl Street served the entire Wall Street and East River area. He did his most important research at a laboratory at Menlo Park, New Jersey, and later West Orange, New Jersey.

Virtually from the moment Tesla stepped into Edison's domain, the company was beset by crises. Fires resulting from faulty electrical wiring threatened the mansions lighted by Edison Electric. The S.S. *Oregon,* the first ocean liner with electric lighting, was delayed at the docks and losing money by the hour because its dynamos were unable to generate power. Edison, eager to expedite repairs so as not to jeopardize ties with his major financial backer, J. Pierpont Morgan, promptly dispatched his newly arrived assistant to make the necessary repairs. Tesla later recounted:

> *In the evening I took the necessary instruments with me and went aboard the vessel where I stayed the night. The dynamos were in bad condition, having several short-circuits and breaks, but with the assistance of the crew I succeeded in putting them in good shape. At five o'clock in the morning, when passing along Fifth Avenue on my way to the shop I met Edison with Batchellor . . . returning home to retire. "Here is our Parisian running around at night," he said. When I told him that I was coming from the* Oregon *and had repaired both machines, he looked at me in silence and walked away without another word. But when he had gone some distance I heard him remark: "Batchellor, this is a d—n good man," and from that time on I had full freedom in directing work.* (My Inventions)

With a promise of $50,000 for designing improved dynamos, Tesla worked on the project every day, seven days a week, until completing the task. When he went to his boss to report his success and claim his payment, Edison dismissed him with a laugh and curt reply, "You don't understand our American humor."

Nor did Tesla ever have the chance to present his alternating-current theories or demonstrate his induction motor to Edison. His every request fell on deaf ears. Edison was heavily invested in delivering direct current to his profitable line of arc-lighting systems. Although Tesla would work for Edison for only six months, their strained relationship would have far-reaching repercussions in the 20th century.

After Edison

With the sting of Edison's rebuke fresh in his ears, Tesla knew he needed to press forward with greater resolve. His time working with Edison enabled him to observe the genius close at hand, but Tesla would challenge his theories and wage a war of words with Edison for years to come. In one newspaper article, he wrote:

> *His [Thomas Edison's] method was inefficient in the extreme, for an immense ground had to be covered to get anything at all unless blind chance intervened and, at first, I was almost a sorry witness of his doings, knowing that just a little theory and calculation would have saved him 90 per cent of the labor. But he had a veritable contempt for book learning and mathematical knowledge, trusting himself entirely to his inventor's instinct and practical American sense. In view of this, the truly prodigious amount of his actual accomplishments is little short of a miracle.*

Bolstered with confidence after his experience with Edison, Tesla set out to organize his own notebooks on advanced arc lighting and the construction of commutators. He reasoned that this would be a valuable first step in advancing his own ideas on alternating current. A scheme took shape.

Tesla was a man of strong opinion and well-formulated ideas. He could visualize the structure of an electrical machine down to the tiniest wire. While working at Edison Machine Works, he also developed a reputation for hard work and innovative problem-solving. That reputation spread beyond the offices of the Edison plant and enabled him to gain access to individuals who might further his plans.

Perhaps it had not occurred to Tesla immediately that he should apply for design patents. In his case, as for other inventors, to patent a process or design does not necessarily require physical creation. One example would be the development of patent leather. This invention was left to Seth Boyden, who in 1888 made

SETH BOYDEN

hard, shiny leather for the manufacturing of boots in Newark, New Jersey. When asked why he didn't request a patent for *patent leather,* Boyden said, "I introduced patent leather, but it should be remembered that there was nothing generous or liberal in its introduction, as I served myself first, and when its novelty had ceased and I had other objects in view, it was a natural course to leave it." Though he did not seek fame or fortune for his invention, Boyden was greatly respected in his own lifetime. He lived out his remaining years in a house in Maplewood (then called Hilton), New Jersey, that had been donated to him by grateful industrialists. In 1926, an admiring Thomas Edison said of Boyden, "He was one of America's greatest inventors.... His many great and practical inventions have been the basis for great industries which give employment to millions of people." Seth Boyden—not Tesla—was the man Edison named as the second-greatest inventor in American history, after himself.

Tesla Electric Light and Manufacturing Company

Soon after he resigned from the Edison organization in 1885, fortune smiled on Tesla. He was approached by two New Jersey residents, Benjamin A. Vail, a lawyer, and Robert Lane, a businessman, who were both excited about the prospects for electric lighting and wanted a piece of the pie. Their plan was to enlist Tesla in a new business enterprise called the Tesla Electric Light and Manufacturing Company. They would join a burgeoning field to make and distribute arc-lighting equipment across the country.

ARC LIGHTING

Lighting is produced when an arc of electricity jumps between two charged conductors. In the case of a light bulb the conductors are connected by a tightly wound tungsten filament that glows between the conductors, thus emitting light from the bulb.

Vail and Lane made Tesla a partner and issued him stock in the company. Buoyed by his newfound status and position, Tesla set to work on improving his designs for manufacturing arc-lighting equipment, generators to power the equipment, and regulators to control the amount of electricity flowing to the equipment. His improvements resulted in more efficient methods to deliver light, with less energy loss and reduced overheating.

Now it was the moment for Tesla to protect his ideas by patenting them. He would have something to show for his work beside his physical exertion. According to biographer Marc J. Seifer, Tesla met in March 1885 with the noted patent attorney Lemuel Serrell, a former agent of Edison's, and Serrell's patent artist, named Raphael Netter. Over the next several months, Serrell and Netter filed a series of patents that were assigned to the Tesla Electric Light and Manufacturing Company in return for stock shares in the company. The initial set of patents reflected major improvements in the components of the arc-lighting process. PATENT 335,786 – ELECTRIC-ARC LAMP was the first of some 300 patents Tesla filed in his lifetime.

PATENT HISTORY

A patent is an exclusive right granted by a government to protect the owner of an idea to manufacture, use, or sell an invention for a certain number of years. A written application with a detailed description and attendant drawings is filed with the United States Government Patent and Trademark Office. Once approved, the inventor's rights are protected. Currently this protection is for 21 years.

In telling fashion, Tesla clung to his European roots in filing the official document with the United States Government, beginning the application as follows:

> *To all whom it may concern:*
> *Be it known that I, NIKOLA TESLA, of Smiljan Lika, border country of Austria-Hungary, have invented . . .*

It would not be until 1891, when he became a naturalized U.S. citizen, that Tesla would exclude the name of his homeland from his patent applications.

It was Tesla's hope that once he filed the patents and demonstrated real improvements in the arc-lighting system, he would be able to win the support of Vail and Lane for his AC motor system. But whenever he broached the subject, his two partners strung him along. Like Edison before them, Vail and Lane were entrepreneurs. They already had a system that worked well and, they believed, optimized their profits. Arc lighting was the talk of Rahway, NJ, where their system illuminated the town's streets and factories. Vail and Lane were not the least bit interested in branching out into another untested project.

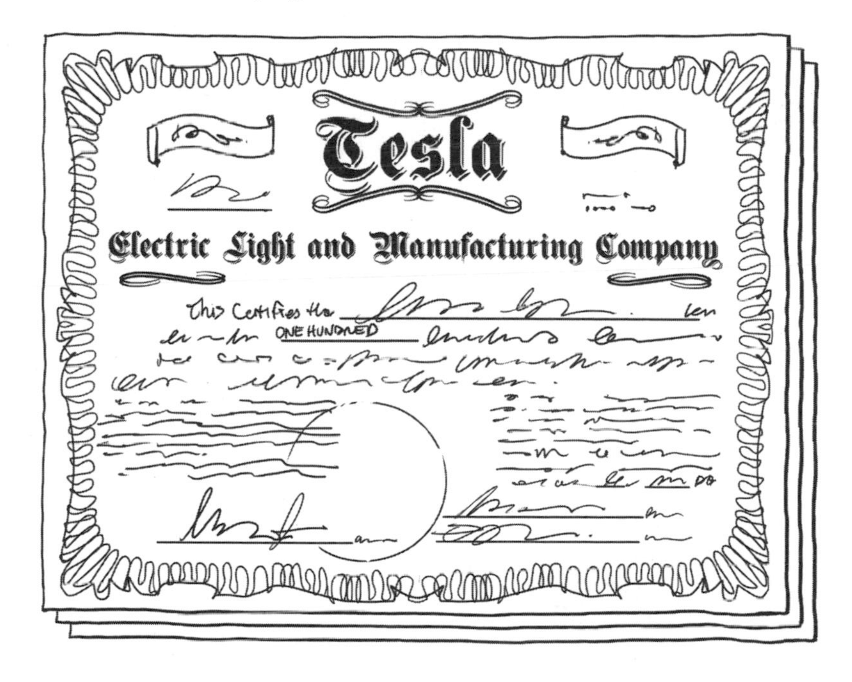

Once Tesla assigned the patents to Tesla Electric Light and Manufacturing Company, the two partners saw no further need to employ their inventor. So they disbanded the company and opened a new one to distribute lighting to the growing city of Rahway. As Tesla recounted it, "In 1886 my system of arc lighting was perfected and adopted for factory and municipal lighting, and I was free, but with no other possession than a beautifully engraved certificate of stock of hypothetical value."

The later part of the 19th century and early part of the 20th century were wild times for inventors, investors, and individuals with business acumen who sought to protect major technological innovations by filing for patents. Great strides in transportation, the transmission of energy, industrial processes, domestic convenience, and home entertainment were spreading across the United States with a rapid rate. Edison had invented the phonograph in 1877 and, with Louis Howard Latimer, he would invent the first long-lasting incandescent light bulb in 1879. Soon to come were the invention of the airplane by Orville and Wilbur Wright in 1903, the Model T by Henry Ford in 1908, and the large-scale moving assembly line in 1913.

LOUIS HOWARD LATIMER (1848–1928)

An expert electrical engineer, Latimer did work for Edison that proved critical to the company's success. His knowledge of electric lighting and power guided Edison through the process of filing patent forms properly at the U.S. Patent Office, protecting the company from infringements of his inventions. Latimer was also in charge of the company library, collecting information from around the world and translating data from French and German to protect the company from European challenges. He became Edison's patent investigator and expert witness in cases against persons trying to benefit from Edison's inventions without legal permission.

Edison encouraged Latimer to write his popular book, *Incandescent Electric Lighting: A Practical Description of the Edison System* (1890), which explained how an incandescent lamp produces light in an easy-to-understand manner. On February 11, 1918, Latimer became one of the 28 charter members of the Edison Pioneers, the only African American in this prestigious group.

Down in the Ditches

Tesla's AC power system would eventually fit neatly into the excitement of the Industrial Revolution and have as great an impact as any of its other inventions. Tesla's arrival in New York in June 1884 came barely a year after the opening of the Brooklyn Bridge, providing road and rail access between Brooklyn and Manhattan. Unlike John and Emily Roebling's "Great Bridge," however, Tesla's flights of fantasy regarding the production, purchase, transmission, distribution, and sale of electricity for residential, commercial, and industrial purposes too often remained brilliant ideas rather than becoming realities.

For all his dreaming, Tesla's concept of an AC power system still had not been developed. And after being abandoned by Vail and Lane, the inventor needed a way to support himself. Now, instead of a laboratory, he found himself literally digging ditches in New York City. The Brooklyn Bridge loomed within his sightlines, glowing after sunset with 70 arc lamps operated by another competitor, the United States Illuminating Company.

The Brooklyn Bridge was the longest suspension span ever built to that time, a testament to modern engineering, entrepreneurial ingenuity, and business acumen. These were three lessons that Tesla had not yet fully grasped. While innovations in electrical engineering still reverberated in his mind, the great practical questions were left unanswered.

What would he need to do in order to convince investors that his ideas were viable? Would Tesla's AC system render all the time, energy, and significant investments in developing, manufacturing, and distribution of electricity via the DC system obsolete? Would he be able to prove that his system would generate large profits? Would the system prove safe? Would it require the circus mentality of P.T. Barnum, who took 21 elephants across the bridge in May 1884, to demonstrate the superiority of AC?

In the meantime, he continued to dig ditches. His savoir faire (sophistication, correct behavior, adaptability) and his ability to speak many languages, including French, did him little good with a shovel in his hand. Nor did his lack of business savvy help the situation. Here he was at the age of 30, by which time many men were well-established in their careers. Tesla had come to America with great hope and promise, but success now seemed farther off than ever and he was plagued by material want, utterly depressed. "My high education in various branches of science, mechanics, and literature," he would later write, "were a mockery."

Tesla Electric Company

With little to show for his efforts other than dirty hands, worthless stock certificates, and copies of patents assigned to other parties, there still had to be a way for him to climb out of the hole. One possibility might be to file a patent in his own name. As in the case of the arc-lamp application, however, the detailed description and diagram could not be dashed off so easily. The application had to be specific to prevent anyone from infringing on the patent. Filing the paperwork was often a long, arduous process, drawing the inventor into an extended love/ hate relationship with the patent office. And there was not even a guarantee that the patent would be granted. How Nikola Tesla had the legal knowledge, skill, and patience to file successfully for PATENT 396,121 – THERMO-MAGNETIC MOTOR, 1886, remains a mystery and a credit to his resourcefulness to the present day.

Although a strong knowledge of electrical engineering is necessary to thoroughly understand many of Tesla's patents, at the heart of this electromagnetism patent is the simple "Right-Hand Rule" of physics as observed by Michael Faraday (1831) in presenting his theory of electromagnetic induction theory. The diagrams below illustrate the Right-Hand Rule.

In the diagram below, imagine a conducting rod placed perpendicular across the index finger with a current flowing through it in the direction of Induced Current I. When a current (in the direction indicated) passes through this rod or the rod in the right picture, a magnetic field radiates out from them. Starting and stopping the flow of current changes the magnetic field—a key principle of Tesla's work on motors.

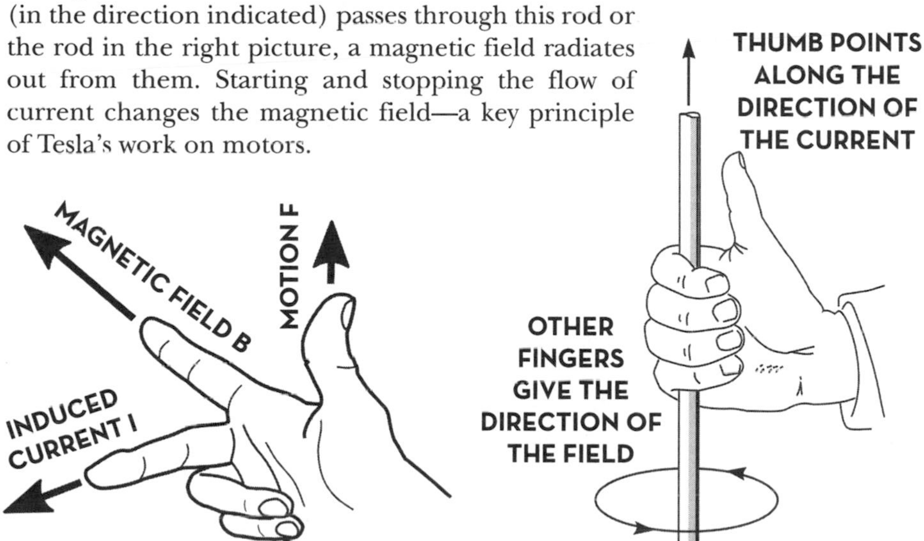

Tesla knew that magnets lose their magnetic strength when heated. To demonstrate this phenomenon, he describes a small motor in his patent application for the thermo-magnetic motor. The motor consists of a fixed magnet (N), an iron pivoting arm (P) with a moving magnet (A) attached to it, a leaf spring (FM), a Bunsen burner (H), and a flywheel. At normal temperature, the fixed magnet is strong enough to pull the pivot arm and compress the spring. But when the pivot arm is pulled toward the fixed magnet, it passes across the flame of the Bunsen burner. The flame heats the pivot arm and causes it to lose the magnetism induced by the fixed magnet. The force of the compressed spring is now greater than the force of the magnetic field, causing the pivot arm to swing away from the fixed magnet. Because the pivot arm is connected by a crank to the flywheel, the motion of the pivot arm causes the flywheel to turn. As the pivot arm swings out of the flame, it cools off and is attracted once again to the magnet. Now the strength of the magnetic field is greater than the force of the spring, causing the pivot arm to swing back toward the fixed magnet and the flame. In short, Tesla's patent application outlined the basic principle of the motor. Beyond that, it was instrumental in opening the door to his development of the AC motor.

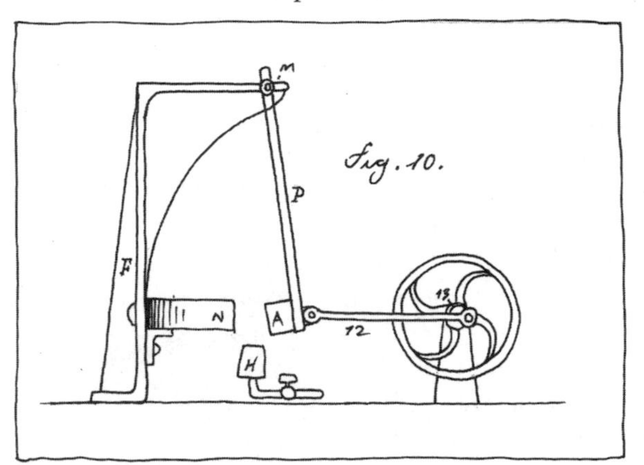

TESLA'S THERMO-MAGNETIC MOTOR

What may have sounded like the ramblings of a mad man digging ditches—talk of lost patents, a lost company, lost inventions, and the genius of his alternating-current system—caught the ear of Tesla's foreman, who introduced him to Alfred S. Brown, a prominent engineer for the Western Union Telegraph Company. Brown himself held a number of patents on arc lamps and was well aware of the limitations of the prevailing DC apparatus. He was immediately impressed with Tesla's ideas and contacted a "distinguished lawyer" from Englewood, New Jersey, named Charles F. Peck.

Brown and Peck were well-versed in business, corporate takeovers, and the lawsuits being filed among the various telegraph companies spreading across the United States. Their experience and practical knowledge in exploiting technological innovation made them ideal partners for Tesla. In Tesla, conversely, Brown and Peck appeared to have found the perfect partner to exploit the new technology.

Peck, however, remained skeptical. He knew that others had failed in creating a viable AC system and did not even want to see a demonstration. In attempting to win his support, Tesla recalled the tale of Christopher Columbus in seeking an audience with the queen of Spain to fund his exploration. In order to do so, he wagered courtiers that he could stand an egg on end. The crowd gathered around and tried in vain. It tried some more and failed. Finally Columbus stepped in, cracked the egg gently on one end, and stood it on the flat surface. He won the bet and secured the necessary funding from Queen Isabella.

Tesla repeated the experiment for Peck, who did not immediately see the connection between Columbus's Egg and Tesla's boasts of creating a thermo-magnetic device, a key element of the AC electrical system. Shrewdly, Tesla challenged Peck to provide funding if he could do better than Columbus and make the egg stand without even cracking the egg. Peck agreed.

Tesla wasted no time over the next few days, enlisting the help of a blacksmith to cast a hard-boiled egg out of copper and brass, and fastening his four-coil magnet to the underside of a wooden table. When he reconvened with Peck and Brown, Tesla placed the copper egg on the top of the table and applied two out-of-phase (or alternating) currents to the magnet. To their astonishment, the egg stood on end. They were even more stupefied when the egg and four brass balls started spinning by themselves on the tabletop. While it looked like magic, Tesla explained, the egg and balls were spinning because of the rotating magnetic field. Peck and Brown were duly impressed by this demonstration and became ardent supporters of Tesla's work on AC motors.

Tesla, for his part, learned two lessons from these demonstrations. First, it is easier to convince someone if they see something with their own eyes. Second, the hardest part is being brave enough to try something new. Showmanship became an integral part of many of Tesla's successes in the future.

Together, Brown and Peck raised the capital and provided the technical expertise to set up Tesla in a laboratory. The facility was located at 89 Liberty Street in Lower Manhattan, next to the grounds on which the World Trade Center would later be built. A partnership with Tesla was formed, and the new company was called the Tesla Electric Company.

Under the requirements of the U.S. Patent Office, a single all-inclusive patent could not cover the entire AC system. The system would have to be broken down into separate groups or components, with individual patents filed for each invention in a particular group. Beginning on April 30, 1887, and over the course of the next year, the Tesla Electric Company engaged in feverish patent filing for an AC dynamo and other devices that comprised Tesla's AC system.

Because his inventions were so new and unique in the burgeoning field of electrical science, they encountered little difficulty in gaining approval. The chart below lists some of the company's key early patents. Repeat titles reflect improvements made in successive filings.

TESLA ELECTRIC COMPANY'S EARLY PATENTS

NO.	TITLE	APPLICATION
382845	Commutator for Dynamo Electrical Machines	April 30, 1887
381968	Electro Magnetic Motor	October 12, 1887
381969	Electro Magnetic Motor	November 30, 1887
382279	Electro Magnetic Motor	November 30, 1887
390414	Dynamo Electric Machine	April 23, 1888
390820	Regulator for Alternate Currrent Motors	April 24, 1888
390721	Dynamo Electric Machine	April 28, 1888
390415	Dynamo Electric Machine for Motor	May 15, 1888

Tesla worked relentlessly, often going without sleep. From memory, he was able to reproduce machines he had conceived more than five years earlier in Europe. He designed and produced complete systems of alternating-current machinery: single-phase, two-phase, and three-phase. For each system, he conceived the dynamo for generating currents, the motor for producing power

TESLA'S POLYPHASE SYSTEM

Polyphase means multiple currents, each out of phase, or out of step, with one another. Think of a bicycle. When the rider pushes down on one pedal, the other pedal crests at the top and starts to push down, powering the bicycle forward. Similarly, in out-of-phase AC current, there is always one cycle nearing its peak, pushing the current forward.

from them, and the transformers for raising and reducing the voltages, as well as a variety of devices for automatically controlling the machinery—including the mathematical theories for all the components.

Nikola Tesla had finally arrived. His patent work in 1887–1888 marked the beginning of a remarkable 15-year run in the field of invention.

Peck and Brown, for their part, developed an astute business strategy, seeking support and endorsements from eminent electrical engineering professionals. Most notably, George Westinghouse, another daring young electrical pioneer and a practical business man, would open his door to back Tesla financially. Thus began the "The War of the Currents" between Tesla's AC system and Edison's DC system.

Chapter 4
SOARING

ON OCTOBER 28, 1886, President Grover Cleveland officiated at the dedication of Frederic Auguste Bartholdi's great statue, "Liberty Enlightening the World"—now widely known as the Statue of Liberty—in New York Harbor. Not far away in Lower Manhattan, Tesla's Liberty Street lab soon became a hotbed of design, patent applications, and the construction of various AC induction motors. It was work that would keep Tesla awake around the clock for days on end.

In May 1887, Tesla recruited his old friend and associate from Budapest, Anthony Szigeti (who had witnessed Tesla's early sand drawings of an induction motor almost five years earlier)

to assist with production. As his ship sailed past the outstretched torch of Lady Liberty, Szigeti's grandest dreams of success could hardly have anticipated what lay ahead. Along with engineer Alfred Brown, who provided technical input, Tesla and Szigeti would forge an alliance to free the world from the constraints of DC.

Now that funds were raised for research and development, it was an opportune time to bring Tesla's induction motor to the attention of the electrical engineering establishment. First, Peck and Brown would need to vet the machine with a prominent electrical engineer and physicist before they could realize any profit. They found the right person in Professor William A. Anthony of Cornell University, who subjected two motors—one AC, one DC—to intense scrutiny. According to Professor Anthony's report, the efficiency of Tesla's AC motor was equal to that of the best DC motor.

Second, Peck and Brown were anxious to reach out to the electrical community at-large. With Professor Anthony's glowing review, it was easy to attract the attention of Thomas Commerford Martin, the editor of America's premier electrical magazine, *Electrical World*, as well as president of the American Institute of Electrical Engineering (AIEE). His endorsement would be key to Tesla's success in America.

Thomas Commeford Martin and the AIEE Lecture

After a demonstration of Tesla's apparatus at the Liberty Street lab, Martin suggested that Tesla prepare a lecture for other noted electricians. Tesla refused at first, but when his first seven patents were granted on May 1, 1888, Martin approached him again.

An Englishman educated as a theologian, Martin came to the United States with Old World manners and a close familiarity with the spread of electricity. He had stood beside his father in the laying of the first telegraph cable under the Atlantic Ocean with British physicist Lord Kelvin and others. Martin was the same age as Tesla and shared a common interest in electrical research. Upon his arrival in New York in 1877 he, too, had gone to work for Edison.

THOMAS COMMEFORD MARTIN

After the demonstration at Liberty Street, Martin was convinced that Tesla's inventions in the utilization of polyphase currents would show how thoroughly the inventor had mastered the fundamental concept and could apply it in the greatest variety of ways. Martin persisted in asking Tesla to deliver a lecture to the community of electrical engineers, knowing it would prove that motors no longer required brushes and commutators.

Despite being overworked and in poor health, Tesla succumbed to Martin's persuasive powers (so much so, in fact, that he eventually gave Martin approval to re-publish all of his papers that had been read before technical societies in the United States and Europe over the next decade). Overnight he wrote out his AIEE lecture in longhand. The following day, May 16, 1888, at Columbia College, he delivered his milestone paper, "A New System of Alternate Current Motors and Transformers."

Imagine the setting: Tesla, dressed impeccably as always, standing tall over two motors set up for demonstration. He speaks perfect English, perhaps with a slight accent (there is no known recording of his voice). Gathered before him sits an assemblage of skeptical, starch-collared electrical engineers. If archival pictures from the period are any indication, they are a formidable lot. Tesla addressed them as follows:

> *The subject which I now have the pleasure of bringing to your notice is a novel system of electric distribution and transmission of power by means of alternating currents, affording peculiar advantages, particularly in the way of motors, which I am confident will at once establish the superior adaptability of these currents to the transmission of power and will show that many results heretofore unattainable can be reached by their use; results which are very much desired in the practical operation of such systems, and which cannot be accomplished by means of continuous currents.*

Tesla went on to explain that commutators are useless in a motor because they unnecessarily reverse electrical currents. He supported his presentation with a series of diagrams detailing the step-by-step, back-and-forth magnetic flow of electricity in his Electro-Magnetic Motor. The patent for it, Number 381,968, had been approved just two weeks prior to the lecture.

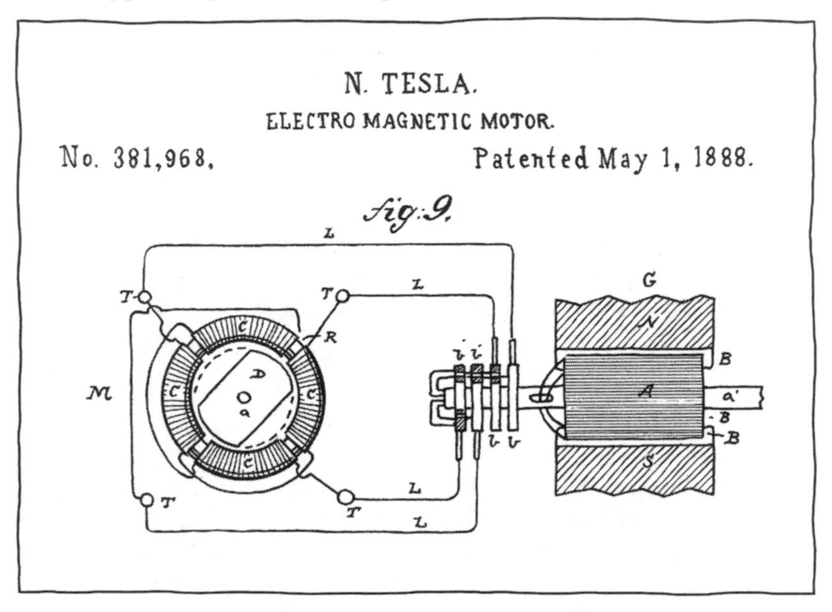

It was a convincing demonstration of how two alternating currents can create a rotating magnetic field and, alternatively, how the rotating magnetic field exerts a uniform pull on the motor's rotor. To be sure, he backed up the demonstration with a full mathematical analysis of how all this was accomplished. He then described a basic polyphase motor, which consisted of a ring with four separate coils for the stator (which remains fixed with respect to rotating parts) and the steel disc rotor. So clear and persuasive was Tesla's lecture that the engineering community was able to absorb the new principles as if they knew his theories all along.

As is often the case with any invention or new theory, however, there is always someone else who claims to have had the idea first or who refuses to let go of established ways of thinking. Tesla's challengers were at the ready, poised to nip away when the floor was opened to discussion.

Professor Elihu Thomson, a prominent electrical engineer and inventor, took the bait. He stood up to make the case against the polyphase motor. Tesla, no one's pushover, responded in a courtly manner by emphasizing the unnecessary use of commutators. Thompson was incensed and remained forever at odds with Nikola Tesla.

While the exchange demonstrated the high degree of competitiveness among electrical engineers and inventors, it served Peck, Brown, and Tesla's desired purpose. The effect was to set the electrical community abuzz, opening the door for the third component of their business strategy: to sell or license the patents.

Enter George Westinghouse

With guile, shrewdness, subterfuge, or smoke and mirrors—call it what you will—Peck and Brown were determined to get the best deal possible for Tesla's invention. Certainly there were profiteers who recognized the potential of owning Tesla's patents. Controlling the distribution of electrical power was irresistible to egomaniacal entrepreneurs, as exemplified by Edison's quest to corner the market on direct current. The rush to control the flow of electricity to homes and factories was in full-throttle. Tesla's system changed the whole playing field.

GEORGE WESTINGHOUSE

George Westinghouse, an industrialist and inventor, had earned a fortune from his invention of railroad air brakes (patented 1869). So successful was this invention that the Railroad Safety Appliance Act of 1893 made air brakes compulsory on all American trains; they also became standard in Europe. Westinghouse also pursued the development of electronic railroad signaling devices and standardization across all railway systems.

For many successful inventors, including Edison, groundbreaking work is predicated on the achievements of others. As George Westinghouse built his empire, he purchased a number of devices and patents from Granville T. Woods, who had filed over 50 patents in his own right but lacked the business savvy or social standing to make the most of them.

GRANVILLE T. WOODS
(1856–1910)

A free black from Ohio, Woods was a self-educated civil and electrical engineer whose inventions significantly enhanced the electrical and communication systems in the railway industry. One of his patents was for a mechanism he called a "telegraphony," a combination telegraph and telephone that could transmit both oral and signal messages. Woods came to be known as the "Black Edison" after successfully defeating Edison's patent challenges.

Westinghouse was no stranger to alternating current and recognized its potential. He had acquired the patents to an early AC distribution system pioneered by the French scientist Lucien Gaulard and English engineer John Dixon Gibbs. With Gaulard-Gibbs transformers and a Siemens AC generator, Westinghouse sought to install an electrical system for the city of Pittsburgh, Pennsylvania. Getting such a system up and running posed enormous challenges. In 1886, he incorporated the Westinghouse Electric Company in direct opposition to Edison's DC system; three years later, he renamed his firm the Westinghouse Electric & Manufacturing Company.

Tesla's AIEE speech captured the attention of George Westinghouse, who directed his staff of engineers to investigate Tesla's motors. Rather than have them try to copy or modify his design, Westinghouse began negotiating with Peck and Brown for the patent rights to Tesla's motor and AC system. He was firm in his conviction that the future lay not only in AC lighting but in AC motors and transformers that could step up and step down the voltage as required. Westinghouse was ready to invest—and heavily—in Tesla's technology. For starts, he had his scientific associates and legal team trying to corner the market on Tesla's motor and patents.

After weeks of negotiation, Peck, Brown, and Tesla were able to secure a favorable deal with Westinghouse for the motor patents. The industrialist agreed to pay $20,000 in cash and $50,000 in Westinghouse stock, plus a $2.50 royalty per horsepower on every Tesla motor; there would be a $5,000 minimum royalty payment in the first year, $10,000 the second year, and $15,000 the third year. Commenting on the deal, Westinghouse said,

> *With reference to the Tesla motor patents, the price to be paid seems rather high when coupled with all of the other terms and conditions, but if it is the only practicable method for operating a motor by the alternating current, and if it is applicable to street car work, we can unquestionably easily get from the users of the apparatus whatever tax is put upon it by the inventors.*

Tesla had great respect for Westinghouse, who, after all, believed in the AC system. With characteristic munificence, Tesla was willing to divide proceeds of the sale, retaining four-ninths for himself while granting five-ninths to Peck and Brown for their efforts.

That summer, Tesla moved to Pittsburgh to help at the Westinghouse plant with the development of motors and the city's power system. With his newfound cash flow, Tesla began his life-long residency in hotels. It was a lifestyle that befitted his status, wealth, temperament, and permanent bachelor status. His trusted friend and assistant Szigeti remained behind to test out new machines in the Liberty Street lab.

Not all of his colleagues at Westinghouse were entirely receptive to Tesla's radical new system. They had long run their motors at 133 cycles per second and resisted Tesla's proposal to run on a frequency of only 60 cycles per second. It was a frustrating period for him, as he wasted valuable time trying to adapt Westinghouse's motors to his proposed frequency. It prevented Tesla from new research for months on end before the Westinghouse staff finally accepted his proposal to run at the lower frequency and make their motors practical. Sixty cycles per second, or 60 Hertz (written as 60 Hz), remains standard to the present today.

In late summer 1889, after a year in Pittsburgh, Tesla returned to New York for a short stay, primarily to set up his second lab at 175 Grand Street. That September, he went to see the recently unveiled Eiffel Tower at the Paris World Exposition, to lecture in Europe, and to visit his family in Serbia.

AC vs. DC: "War of the Currents"

By acquiring the Tesla AC motor, Westinghouse was positioning himself for success on a large scale. Edison was rightly nervous that this posed a major threat to his business built on DC. The result was the so-called War of the Currents. Edison fired the first salvo by initiating a propaganda battle, citing safety concerns over the high voltage utilized in AC systems. Demonstrations were organized to dissuade users from adopting alternating current over Edison's tried-and-true direct current. Edison would not hold back in trying to protect his technology and his profits, going so far as to suggest that electricity—AC in particular—kills. He had invested enormous sums of money to build DC generating stations and networks of copper wire every few city blocks, covering about a half-mile, to deliver electricity underground. By 1890, however, that distance paled by comparison to Westinghouse's installation of a 12-mile, 4000-volt AC transmission line from Willamette Falls to Portland, Oregon.

Enter Harold P. Brown, a New York engineer and electrical consultant who endeared himself to Edison's cause by publishing articles on the dubious merits of AC. Brown went so far as to offer 25 cents for every dog and cat rounded up for public electrocution by being wired to an AC generating motor. The demonstrations would start by wiring a dog to a DC generator and slowly raising the voltage, from a low point that caused the dog mild annoyance to a high point, close to 700 volts, at which the animal broke loose from the wire restraints. The upper level was not enough to kill the dog but still tortuous, and onlookers denounced the exercise. Brown used that moment to further his anti-AC crusade, arguing that that he could put the dog out of its misery by switching to alternating current. As he demonstrated on stage, a quick jolt of 300-volts of AC brought about the quick death of the dog. Had the initial DC charges so thoroughly weakened the animal that it only took the final AC charge? In reality the demonstration did not prove that AC was any more lethal than DC, but it strengthened that perception and helped affirm Edison's position in the public's eye.

Although it is not completely clear what motivated Brown to electrocute dogs and cats, we do know that he was employed by Edison and that he was seeking a more humane method of executing violent criminals. Edison himself later attended hearings that spurred the New York State Legislature to adopt the electric chair for capital punishment. Westinghouse was outraged, regarding the campaign as part of the propaganda war against AC.

With Edison's endorsement, Brown was appointed as New York's electrocution expert and wasted little time in having a Westinghouse generator installed at Auburn State Prison. The first legal execution by electrocution took place there on August 6, 1890, with Brown in attendance. The execution did not go at all smoothly. After a succession of clumsy attempts, convicted murderer William Kemmler finally succumbed to increased dosages of AC electricity. It was a grisly affair, roundly denounced in newspapers the next day.

Proponents of DC, missing no opportunity to promote their cause, took to saying that Kemmler had been "Westinghoused." George Westinghouse himself said, "I do not care to talk about it. It has been a brutal affair. They could have done better with an axe." Tesla, an ardent humanitarian, in later years expressed his abhorrence of electrical execution. Harold Brown, with all the brouhaha over his demonstrations, quietly disappeared from the War of the Currents. And Edison, still seething from his battle against Westinghouse, continued to discredit his adversary for years. In one last jab, in January 1903, Edison filmed a six-ton elephant named Topsy being publicly electrocuted at Coney Island in New York. By that time, the War of the Currents was long over and Tesla's AC system had been accepted worldwide.

Paris and Radio Discoveries

In the meantime, Edison's early campaign against alternating current rocked Westinghouse's investors. Production of Tesla's induction motors ground to a standstill in 1899 and would not continue until after Tesla returned from Paris.

While work had stopped in Pittsburgh, the Grand Street lab was abuzz. Before departing for Paris, Tesla had begun early experiments in high-frequency apparatus, the relationship between electromagnetic radiation and light, and what would become his lifelong obsession, the wireless transmission of energy.

Barely five years after arriving in the United States with four cents in his pocket, Tesla at age 33 was a wealthy man and could sail to Paris in style. The great attraction there was the Exposition

Universelle, or Paris World's Fair, and its controversial modernist centerpiece, the Eiffel Tower. Also on display at the exposition would be a number of new electrical exhibits and inventions, including the introduction of an Edison phonograph for the general public. While Edison was being lauded and celebrated by European dignitaries, Tesla had other things on his mind. He hoped, for example, that the Exposition might be an opportunity for him to catch up on the latest news in the field of electrical vibration regarding the relationship between alternating current and light waves. Here was his chance to confer with the most advanced theorists in Europe.

During his stay in Paris, Tesla was also able to meet with the young Norwegian mathematician and physicist Vilhelm Bjerknes. It was an auspicious meeting, as Bjerknes had collaborated with French mathematician Jules-Henri Poincaré in reproducing the experiments and had worked out the mathematics that confirmed the findings of German physicist Heinrich Hertz on the propagation of electromagnetic waves through space. For Tesla, the visit was an opportunity to study Bjerknes's oscillator, which provided a variety of electromagnetic waves and a resonator for augmenting them. Further, he discussed theoretical implications concerning the properties of the resulting electromagnetic waves. From these discussions, Tesla discovered that the electromagnetic waves, or Hertzian waves, not only produce transverse oscillations, but also longitudinal vibrations structured much like sound waves. According to Tesla biographer Marc J. Seifer, these findings would be instrumental in Tesla's work over the next decade in his construction of wireless transmitters. It would lead to one of his most important discoveries: the radio.

Chapter 5

THE "BIG SHOW"

FOLLOWING A SERIES OF MEETINGS WITH PROMINENT SCIENTISTS in Europe and a visit with his family, Tesla returned to New York. With the departure of his friend Anthony Szigeti, the small size of his Grand Street laboratory, and his desire for more space, Tesla opened a more spacious facility at 33-35 South Fifth Avenue (now called LaGuardia Place) in 1899. Ironically the lab was located only blocks south of the Edison Electric Light Company's elaborate showroom and headquarters at 65 Fifth Avenue and 14th Street, where Edison's new products were on display.

Back in the U.S.A.

Tesla was buoyed by his European sojourn. He stood out with his elegant attire, white gloves, and Continental manner. He took up residence in one of the country's most prestigious hotels, the Astor House, in Lower Manhattan near St. Paul's Chapel. Now a man of means, he dined regularly at Delmonico's, the most famous and exquisite restaurant in America. But he ate alone, indulging his idiosyncrasies. Phobic about germs, he would

use two dozen cloth napkins in a single meal (adhering to his penchant for items divisible by three), wiping clean his dishes and silverware prior to eating.

Although his fame was spreading, he remained a loner. On occasion he would meet with Thomas Commerford Martin, by now his trusted publicist, or the engineers at Westinghouse who were working on projects associated with his patents. He was especially eager to apply the knowledge he had gained from his meeting with Vilhelm Bjerknes in Paris to his own research on wireless transmission. He plunged into his work, keeping a nocturnal schedule. His single-mindedness and aloofness become the stuff of legend. Transmission expert Frank Jenkins of the Duke Power Engineering Society remembered him like this:

> *He was not an easy person to work for because he expected almost as much from his help as he did from himself. Since he could depend entirely on his memory for all details, he believed others could too if they tried hard enough. He required his machinist to work from memory. He was unable throughout his life to cooperate with others in acquiring knowledge and conducting research ... but preferred to work independently.*

Generosity to Westinghouse

Before he could become completely absorbed in wireless transmission, other problems that demanded Tesla's attention were looming on the horizon. For Westinghouse to continue implementing Tesla's AC system and motor, larger sums of capital would be required. Edison's propaganda campaign was in full swing now, and it had a ripple effect on Westinghouse's own backers. A number of them began to balk. Finally they delivered an ultimatum. Tesla had to modify Westinghouse's equipment to run on higher frequencies, and Tesla's royalty agreement had to be canceled. The demands left Westinghouse in a bind. To convince Tesla to nullify the contract would seriously jeopardize any prospects for Westinghouse to reap the benefits of an AC system in the future. As an inventor himself, moreover, he felt a strong sense of loyalty toward Tesla and understood the value of patent royalties.

Faced with the unpleasant task of confronting Tesla, Westinghouse diplomatically suggested the following: if Tesla did not relinquish the royalty clause, then he (Westinghouse) would lose control over Tesla's patents. The negotiations between the two men were vividly described by Tesla's personal friend and biographer John J. O'Neill:

"I believe your polyphase system is the greatest discovery in the field of electricity," Westinghouse explained....

"Mr. Westinghouse," said Tesla, "you have been my friend, you believed in me when others had no faith; you were brave enough to go ahead and pay me a million dollars when others lacked courage.... The benefits that will come to civilization from my polyphase system mean more to me than money involved. Mr. Westinghouse, you will save your company so that you can develop my inventions. Here is your contract and here is my contract – I will tear both of them to pieces and you will no longer have any troubles from my royalties."

Matching his actions to his words Tesla tore up the contract.... Westinghouse, thanks to Tesla's magnificent gesture, was able to make good his promise to Tesla to make his alternating current system available to the world. (*O'Neill,* Prodigal Genius: The Life of Nikola Tesla)

At $2.50 per horsepower, Tesla lost an estimated $12 million with his act of generosity under this arrangement.

Telluride

Westinghouse was able to begin making good on his promise. The Gold King Mine in Telluride, Colorado, was in desperate financial straits, having all but exhausted its means of powering mining activities. It had depleted neighboring forestland and could no longer afford to cart in coal to fuel its machines. Company manager L.L. Nunn had read about the successes of Nikola Tesla and George Westinghouse with alternating current power and was impressed by their claims that AC could be transmitted much longer distances than DC.

Nunn was able to strike a deal with Tesla and Westinghouse to build the world's first commercial-grade alternating-current power plant in Ames, Colorado, near Telluride. Whereas electricity

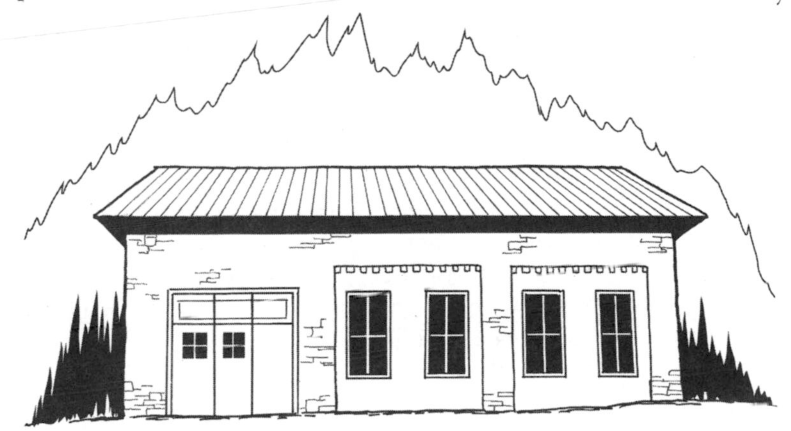

The first electric power plant to produce and transmit alternating current for industrial purposes.

had been used in the past chiefly to power public lighting, the initial transmission line for the Colorado mine delivered enough power to operate a 100-horsepower synchronous motor, itself started by a Tesla AC induction engine. The Ames Power Plant began operation in 1891, and Nunn's venture would provide a model for the development of power distribution at Niagara Falls. For the time being, the word was out—*AC power worked!* It could be transmitted over many miles and power large equipment. The mine and the economy of Telluride were saved.

SYNCHRONOUS MOTOR

A one-speed motor, or fixed motor, in which the speed is governed or synchronized with the supply frequency. In this case, a Tesla AC induction motor supplied the electricity which caused the speed of the rotor to be in sync with the rotating magnetic field of the Tesla motor.

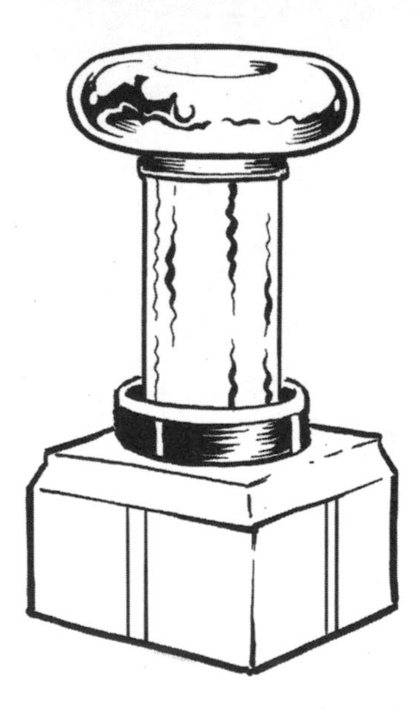

Tesla Coils

In 1890–1891, as Westinghouse was negotiating the deal with Nunn that led to the Ames plant, Tesla was hardly sitting idle. Since he had ceded his royalties to Westinghouse, he would not receive income from the success of the AC installation in Colorado. He would need to develop other sources of revenue, and his mind took off in three major directions. First, having successfully delivered his polyphase induction system, he would seek the means to deliver power wirelessly. Second, he would try to invent new methods of illumination. And third, he would investigate the wireless transmission of intelligence. These broad goals established the direction of work that would occupy Tesla for the next 50 years.

The pursuit of the wireless transmission began at his Grand Street lab as he was busy exploring Hertz's work on high-frequency electromagnetic waves. Tesla was determined to increase the frequency of electrical vibrations until they were equal to light. He hoped to produce light more efficiently than the wasteful process used in Edison's incandescent light bulb. In the delivery of electricity to an Edison bulb, only 5% of the power was effective; the other 95% was lost to heat waves.

CONDENSER OR CAPACITOR

An electric circuit element used to store charge temporarily. A condenser generally consists of two metallic plates that are separated and insulated from each other by a dielectric material, such as porcelain (ceramic).

A dielectric is a poor conductor of electricity, but it is an efficient supporter of an electromagnetic field. When repeated charges are accelerated back and forth through the plates one magnetic field after another is produced. These fields will spread over large distances through space. The result is that capacitors are integral to radio frequencies.

Drawing upon Lord Kelvin's 1856 theory of condenser discharge, for which no apparent use had yet been developed, Tesla embarked upon his own production of electrical oscillations. Kelvin had theorized that when a condenser is discharged, its action is like the up-and-down bobbing that takes place when a weighted, stretched spring is released. The electricity, he showed, rushes from one plate into the other and then back again; the process continues until all the stored energy is used up in overcoming frictional losses. The back-and-forth surges take place at an extremely high frequency, hundreds of millions per second.

Tesla reasoned that if he could build a device that continuously emits electricity at higher frequencies, he would be able to achieve important technical advantages. Among these would be lamps that glow more brilliantly and the ability to transmit energy more efficiently. Tesla began his high-frequency investigations by replicating Hertz's apparatus for giving off an electric spark and propagating electromagnetic waves in space.

At the heart of Hertz's machine was a large induction coil, consisting of an iron core wound with two different thicknesses of wire. A

battery with one lead would be connected to the iron core, and the other lead would run to a telegraph key. That, in turn, would be connected to the thicker of the two wires (primary). Each of two leads—one running from the thinner wire (secondary) and one running from the iron core—would be attached to each of two electrodes, with a gap separating them. Every time current flowing from the battery to the telegraph key was turned on or off, the induction coil produced high-voltage sparks between the two electrodes.

Into this apparatus, Tesla introduced a capacitor and took to manipulating the induction coil in and out. Eventually he would remove the iron core and rely on air cores for separate windings of both the primary and secondary wires, which were tuned to resonate. He would also replace the battery with an AC generator, as well as a step-up transformer.

Sending electricity a long distance would require higher voltage, basically to push the current through the wires. However, the use of high voltages in homes and factories would be dangerous. The step-up transformer made it possible to increase low-voltage, high-current AC to high-voltage, low-current AC at higher frequencies. This was not possible with direct current, however, since DC could only be distributed over short distances. Eventually a distribution transformer was introduced in the system to step down the higher voltages needed in factories, or for high-speed electric trains. For more modest domestic uses, a home-supply transformer would be introduced to further step down the distribution to 110 volts.

Tesla was attempting to manipulate the vibrations of every capacitor discharge and obtain higher voltages from the current produced by an induction coil. Tinkering in this way finally enabled him to control the amount of discharge as well as the frequency. He called his machine an *oscillating generator.* His findings were reflected in his application for Tesla PATENT 462,418 – METHOD OF AND APPARATUS FOR ELECTRICAL

CONVERSION AND DISTRIBUTION, filed on February 4, 1891. It was granted on November 3, 1891. The invention would ultimately become known as the ***Tesla coil.*** With its primary and secondary circuits both tuned to vibrate in harmony, the Tesla coil would be one of his greatest inventions. It would be used in various forms for the future of radio and television.

"Awesome" could easily be the word for what Tesla observed next. Sparks of great length and frequency would spout from the coils in his lab not connected to the power-generating primary coil. Because the different secondary coils were tuned to the same wavelength or frequency of the primary coil, the secondary coils would erupt in sparks. Observing this, Tesla devised all shapes of coils that set the lab awash in streaking jolts of electricity. In working with such high voltages, he made sure to keep one hand in his pocket to avoid completing a circuit that could electrocute him on the spot. At last he was transmitting energy wirelessly! He planned a spectacular demonstration of the new principle, in which he would wave unconnected glass tubes that looked like flaming swords and could, to everyone's amazement, light up a room.

WILLIAM CROOKES

The Crookes radiometer loomed large in Tesla's thinking. English physicist and chemist Sir William Crookes's experiments and spiritual investigations in the 19th century greatly influenced Tesla's thinking. The Crookes radiometer, today considered a novelty item, was basically a glass globe with the air removed. Inside was a system of vanes, each blackened on one side and polished on the other, that rotated when exposed to radiant energy. Crookes developed the radiometer to utilize vacuum balance in his research on the element thallium. He never did provide a true explanation for the apparent "attraction and repulsion resulting from radiation."

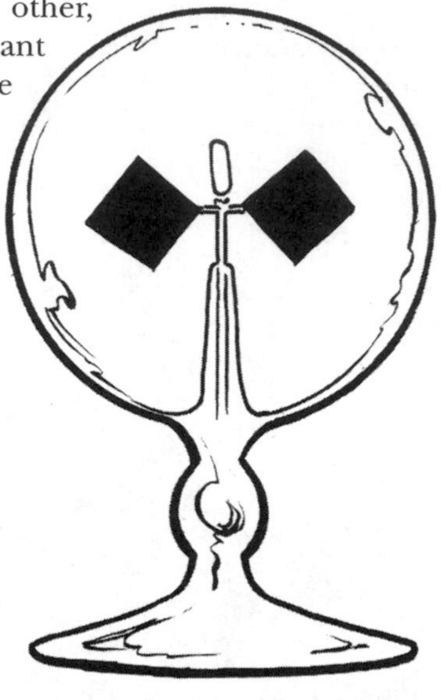

CROOKES'S RADIOMETER

What might have attracted Tesla to the radiometer was Crookes's use early on of high-voltage current produced by an old-fashioned induction coil to achieve rotation. Tesla believed that he could achieve more far-reaching results by fashioning other devices for illumination: glass tubes that enclosed different gases such as argon and xenon, or had the air removed from them; an incandescent carbon button lamp (his single-electrode incandescent lamp); fluorescent tubes, and neon-filled tubes in the shape of letters.

Experimenting in this way, Tesla produced four new kinds of lamps:

1. Tubes in which a solid body became incandescent;
2. Tubes in which phosphorescent and fluorescent materials were made to illuminate;
3. Tubes in which rarefied gases became luminous, and
4. Tubes in which luminosity was produced in gases at ordinary pressure.

Applying the strategy he learned from Peck and Brown, Tesla filed patent applications 454,622 – SYSTEM OF ELECTRIC LIGHTING, and 455,069 – ELECTRIC INCANDESCENT LAMP, in spring 1891. With that done, he was ready to demonstrate his endeavors publicly. He enlisted his friend Thomas Commerford Martin to launch a campaign to publicize his work on high-frequency AC.

Columbia College Lecture

Martin arranged for Tesla to deliver another lecture to the American Institute of Electrical Engineers (AIEE), on May 20, 1891, at Columbia College. Before the same body three years earlier, he had delivered his groundbreaking paper, "A New System of Alternate Current Motors and Transformers." This time he would be an even greater showman, beguiling his audience with fantastic, never-before-seen demonstrations of wireless illumination. To say that his display was mesmerizing would not do it justice. Was he using sleight of hand? Were there tricks up his sleeve? Quite the contrary. He was able to explain each of his demonstrations with prodigious mathematical formulas and detailed schematics. It was all science, and Tesla held his audience in thrall for over three hours.

Biographer Thomas Commerford Martin recorded this lecture and others in *The Inventions, Researches and Writings of Nikola Tesla* (1894). According to Martin's account,

> *Mr. Tesla did not hesitate to show many new and brilliant experiments, and to advance the frontier of discovery far beyond any point he had theretofore marked publicly. . . The ground covered by them is so vast that only the leading ideas and experiments can here be touched upon; besides, it is preferable that the lectures should be carefully gone over for their own sake, it being more than likely that each student will discover a new beauty or stimulus in them.*

THE TESLA LECTURES

"Experiments with Alternate Currents of Very High Frequency, and Their Application to Methods of Artificial Illumination"	AIEE, Columbia College, New York	May 1891
"Experiments with Alternate Currents of High Potential and High Frequency"	Institution of Electrical of Engineers and the Royal Institution of Great Britain, London	February 1892
"Experiments with Alternate Currents of High Potential and High Frequency"	Society of Physics and International Society of Electricians, Paris	February 1892

For the Columbia College lecture, he pulled out all the stops. Not only would he wave around his "flaming swords," he would treat his audience to something very special. Witnesses would report:

> *With hundreds of thousands of volts of high-frequency currents surging across his body, he would hold in his hand a strangely powerful little "carbon-button" lamp. Energy from his body caused gas molecules in the tube to bombard a small button of carborundum until it glowed to incandescence, resulting in a light twenty times brighter than any other lamp in existence. The energy inside the bulb was so powerful that it vaporized diamonds and rubies.* (*Cheney and Uth,* Tesla: Master of Lightning)

For dramatic effect, he held the carbon button lamp aloft like the Statue of Liberty. Although never produced or marketed for practical purposes, the carbon button lamp was used for experiments in which Tesla bombarded rubies with electrical or molecular energy—a forerunner of today's laser technology and atom smashing.

The Columbia lecture produced a groundswell of excitement and a rush to capitalize on his inventions. Still, the naysayers, doubters, and self-promoters came forward to dispute Tesla's findings or to claim credit for having covered the same ground in their own experiments. Tesla took on all challengers, but such prominent electrical engineers as Elihu Thompson and Michael Pupin (a fellow Serb) took the bait and insisted that they had discovered the high-frequency principles long before Tesla. In the end, Tesla's patents affirmed time after time that his work was original and path-breaking.

The full import and possibilities of the Tesla coil have remained undetermined to the present day, since Tesla worked out much of the mathematics in his head, never committing it to paper. Scientists, physicists, and electrical engineers continue in their efforts to unravel many of the secrets Tesla took to his grave. His investigations and trials with the Tesla coil is some of his least understood work.

Similarly, Tesla's work on high-frequency currents led to a number of discoveries that he never patented. Among the by-products are modern electric clocks, x-ray photography, and therapeutic deep-heat healing effects that laid the foundation for the medical field of diathermy. Yet even these advances took a backseat to his lifelong obsession with the wireless transmission of energy. Indeed it was his

work in this area that ultimately earned him the labels of magician, sorcerer, wizard, shaman, mystic, "man with lightning in his hand," "master of lightning," and "electrical messiah."

Citizenship

Just seven years after his arrival in the United States, Tesla's Columbia College lecture catapulted him to the forefront of this country's electrical inventors. With all the accolades and press coverage that followed his presentation to the AIEE, he was pressured to deliver lectures in Europe. Before doing so, however, he achieved what he considered one of his highest honors—U.S. citizenship. "The papers ... that conferred on me the honor of American citizenship," he would later write, "are always kept in a safe, while my other orders, diplomas, degrees, gold medals and other distinctions are packed away in old trunks."

Ironically, Tesla's naturalization documents, dated July 30, 1891, listed him as residing at the Hotel Gerlach (currently the Radio Wave Building), at 49 West 27th Street in New York City. It was an early indication of his movements from hotel to hotel—a pattern that would continue for the rest of his life. He filled in his occupation as "civil engineer" and his former nationality as "Austrian".

The pride Tesla took in his citizenship would be reflected just two days later, on August 1, 1891, when he filed his application for Patent 464,667 – ELECTRIC CONDENSER. In it he proudly identified himself as a "citizen of the United States."

With that patent, Tesla set the record straight regarding improvements above and beyond his demonstration for the AIEE in which he passed gas-filled tubes that glowed between two electrically charged plates. Indeed it marked an important step forward in his efforts to transmit wireless energy over long distances.

London and Paris Lectures

Never one to gloat over the success of his lectures, Tesla was anxious to resume his research on wireless energy after the AIEE event. How currents pass through the Earth was a phenomenon he was especially eager to explore, as highlighted in his later account of the Columbia lecture.

While the spontaneous success of my lecture was due to spectacular features, its chief import was in showing that all kinds of devices could be operated thru a single wire without return. This was the initial step in the evolution of my wireless system. The idea presented itself to me that it might be possible, under observance of proper conditions of resonance, to transmit electric energy thru the earth, thus dispensing with all artificial conductors. (*"The True Wireless,"* Electrical Experimenter, *May 1919*)

Tesla did his best to shut out the many invitations, honors, and other flattering inducements being offered to him and ardently pursued his research in grounding. But the demands for personal appearances finally became irresistible, and Tesla agreed to address the Institution of Electrical Engineers in London.

On the night of February 3, 1892, he took the podium before a British audience in formal attire and rapt attention. Elegant as ever himself, Tesla rose to new levels of showmanship. His "magic wands," button lamp, and gales of electrical sparks circling his body brought resounding cheers from the audience. His coup de grace was to take an apparatus similar to the Crookes radiometer and place it in the electric field generated between two electrified zinc plates, causing the vanes to begin spinning. Here was Tesla demonstrating the first wireless motor.

On hand for Tesla's demonstration were some of England's most prominent physicists and electrical engineers, including his heroes William Thomson (Lord Kelvin) and Sir William Crookes. So ecstatic was the crowd that he was urged to repeat his lecture the following evening for an equally august body of leading scientists. Sir James Dewar, a Scottish chemist and physicist best known for his work with low-temperature phenomena and the inventor of the Dewar flask, was instrumental in bringing about the second lecture. Recounting his reluctance to stay on, Tesla wrote the following:

It had been my intention to leave immediately for Paris in compliance with a similar obligation, but Sir James Dewar insisted on my appearing before the Royal Institution. I was a man of firm resolve but succumbed easily to the forceful arguments of the great Scotchman. He pushed me into a chair and poured out half a glass of a wonderful brown fluid which sparkled in all sorts of iridescent colors and tasted like nectar. "Now," said he, "you are sitting in Faraday's chair and you are enjoying whiskey he used to drink." In both aspects it was an enviable experience. The next evening I gave a demonstration before that Institution. (My Inventions)

His stay in London also enabled Tesla to solidify his European electrical patents and to meet with many other physicists and electrical engineers. Most notable for him was the opportunity to present Lord Kelvin and Sir William Crookes with their own Tesla coils. His extended visit with Crookes was devoted to discussing their respective research, the potential uses of electricity, their personal backgrounds, and their religious beliefs. The discussion turned to the supernatural, of which Tesla was not an actual believer. Exhausted and under strain, he was not able to muster much of a defense against Crookes, who would later serve as president of the Society of Psychical Research. Indeed Tesla paused to consider. "I might not have paid attention to other men," he later wrote, "but was susceptible to his [Crookes's] arguments, as it was his epochal work on radiant matter, which I had read as a student, that made me embrace the electrical career. I reflected that the conditions for a look into the beyond were most favorable."

From London he rushed to Paris, where he was to deliver two scheduled lectures. To say that Tesla's lectures and demonstrations rocked the news media and scientific community would be an understatement. After the first of the two lectures in Paris, however, he returned to his hotel and received a message that his mother Djuka was in failing health back in Gospić. Tesla would not stay around to give the second lecture or to bask in all the attention, even if that was his style. The sheer effort of lecturing, answering questions, and dealing with patent issues, combined with his irregular sleeping habits, was catching up with him. He left Paris immediately.

Staying on in Europe

The sleepless rush to Gospić caused a patch of Tesla's hair to turn white overnight, though it returned to its natural jet black color within a month. In any event, he arrived at his dying mother's bedside just in time.

Tesla never heeded the cautions of Sir William Crookes and others that he seemed on the verge of a physical or nervous breakdown. His condition at this time led to a harrowing experience that he recounted years later:

> *I had become completely exhausted by pain and long vigilance.... As I lay helpless, I thought if my mother died while I was away from her bedside she would surely give me a sign.... I reflected that the conditions for a look into the beyond were most favorable....*

During the whole night every fiber in my brain was strained in expectancy, but nothing happened until early in the morning, when I fell in a sleep … and saw a cloud carrying angelic figures … one of whom gazed upon me lovingly and gradually assumed the features of my mother. The appearance slowly floated across the room and vanished, and I was awakened by an indescribably sweet song of many voices. In that instant a certitude, which no words can express, came upon me that my mother had just died.… This occurred long ago, and I have never had the faintest reason since to change my views on psychical and spiritual phenomena, for which there is absolutely no foundation. (My Inventions)

This would not be Tesla's first nor last journey into the psychic or supernatural world. Crookes's admonitions and own spiritual research were intrinsic to a great deal of Tesla's thinking thereafter, and ultimately would play a large part in his unraveling.

Following his mother's passing on April 4, 1892, Tesla suffered the predicted breakdown. He spent a number of weeks recovering and visiting with grieving family members. During the time in his home country, he was accorded special honors by various Serbian dignitaries and solidified his status as a national hero. Yet it was also a time of mental stocktaking, since he was unable to conduct any laboratory work and needed to restore his own energy.

Shaping a Grandiose Plan

Finally it was time to formulate plans and resume his work in America. With his health fully restored, Tesla mulled over advice he had been given in London by Lord Rayleigh:

I never realized that I possessed any particular gift of discovery but Lord Rayleigh, whom I always considered as an ideal man of science, had said so and if that was the case I felt that I should concentrate on some big idea.

One day, as I was roaming in the mountains, I sought shelter from an approaching storm.... All of a sudden, there was a lightning flash and a few moments after a deluge. ... [T]he two phenomena were closely related, as cause and effect, and a little reflection led me to the conclusion that the electrical energy involved in the precipitation of the water was inconsiderable, the function of lightning being much like that of a sensitive trigger.... If we could produce electric effects of the required quality, this whole planet and conditions of existence on it could be transformed. The sun raises the water of the oceans and winds drive it to distant regions where it remains in a state of most delicate balance. If it were in our power to upset it . . . this mighty life-sustaining stream could be at will controlled. We could irrigate arid deserts, create lakes and rivers and provide motive power in unlimited amounts. This would be the most efficient way of harnessing the sun to the uses of man. The consummation depended on our ability to develop electric forces of the order of those in nature. It seemed a hopeless undertaking, but I made up my mind to try it and immediately on my return to the United States, in the summer of 1892, work was begun ... because a means of the same kind was necessary for the successful transmission of energy without wires. (My Inventions)

Tesla's time in Europe would prove pivotal. He set the English and French scientific communities abuzz, wrapped up family matters, and was ready to return to the United States with renewed vigor. Upon his arrival in New York, he settled into the Hotel Gerlach and increased his Fifth Avenue laboratory staff. He was ready to embark on global and perhaps interstellar research, as suggested by his discussions with Rayleigh and Crookes.

This time, however, he was interrupted by two lectures to which he had committed, at the Franklin Institute in Philadelphia in February 1893, and before the annual meeting of the National Electric Light Association in St. Louis in March 1893. Although much of the material was similar to his previous lectures, for the first time he expanded on the tuning of two different coils to the same frequency. In essence this was a rudimentary display of the essential components of radio: a transmitter, receiver, antenna, ground connection, and tuning device. The spark of one coil would transmit electric waves through the air that had been received by the other coil and then converted back into electricity. In his lecture at the Franklin Institute, titled "High Frequency and High Potential Currents," Tesla told the audience:

> *I would say a few words on a subject which constantly fills my thoughts and which concerns the welfare of all. I mean the transmission of intelligible signals or perhaps even power to any distance without the use of wires. I am becoming daily more convinced of the practicality of the scheme.... I know full well that the great majority of scientific men will not believe that such results can be practically and immediately realized.... My conviction has grown so strong, that I no longer look upon this plan of energy or intelligence transmission as mere theoretical possibility, but as a serious problem in electrical engineering, which must be carried out some day. The idea of transmitting intelligence without wires is the natural outcome of the most recent results of electrical investigation.*

As if these words were not enough to strike awe in his listeners, he brought them to the edge of their seats with a demonstration he described and explained in the lecture as follows:

> *On one set of the terminals of the coil, I have placed a large sphere of sheet brass, which is connected to a larger insulated brass plate.... I now set the coil to work and approach the free terminal with a metallic object held in my hand.... As I approach the metallic object to a distance of eight or ten inches, a torrent of furious sparks breaks forth from the end of the secondary wire, which passes through the rubber column. The sparks cease when the metal in hand touches the wire. My arm is now traversed by a powerful electric current, vibrating at about the rate of one million times a second. All around me the electrostatic force makes itself felt, and the air molecules and particles of dust flying about are acted upon and are hammering violently against my body. So great is this agitation of the particles, that when the lights are turned out you may see streams of feeble light appear on some parts of my body.... The streamers offer no particular inconvenience, except that in the ends of the fingertips a burning sensation is felt.... The streams of light which you have observed issuing from my hand are due to a potential of about 200,000 volts, alternating in rather irregular intervals, sometimes like a million times a second.*

Years later, he would proudly sit in front of a camera to pose for the often reproduced photo of him holding a wirelessly illuminated light bulb—confirmation of his unofficial titles as "Father of the Wireless."

World's Columbian Exposition

During Tesla's absence overseas, Westinghouse faced dire financial problems. He found it financially burdensome to develop Tesla's polyphase AC system and was more interested in broadening the single-phase AC system, for which there was an existing market.

Westinghouse was on the verge of losing his status in the electricity industry, as larger electric companies were swallowing up smaller companies. Even Edison General Electric and the Thomson Houston Electric Company had merged in February 1892 to become General Electric (GE). J.P. Morgan, one of Edison's principal investors, attained even greater wealth and financial control with the merger. All but forgotten in the takeover was the fact that Morgan had been the first person to install Edison's incandescent lighting system in his home. Westinghouse, his credit overextended, needed to do something bold to regain credibility with his investors.

Looming on the horizon was the World's Columbian Exposition—the Chicago World Fair of 1893—to mark the 400th anniversary of Christopher Columbus's arrival in the New World. Frederick Law Olmsted, America's foremost landscape architect, was responsible for laying out the fairgrounds, which occupied 639 acres next to Lake Michigan. It took three years to construct the fairgrounds and its 200 buildings. The exposition featured more than 250,000 displays and attractions, including the world's first Ferris Wheel, invented by George W. Ferris. Sophie Hayden, the first woman awarded a degree in architecture from the Massachusetts Institute of Technology (MIT), designed the famous Woman's Building. Only one structure from the original fairgrounds still stands today: the Palace of Fine Arts, transformed in 1931 into the Museum of Science and Industry.

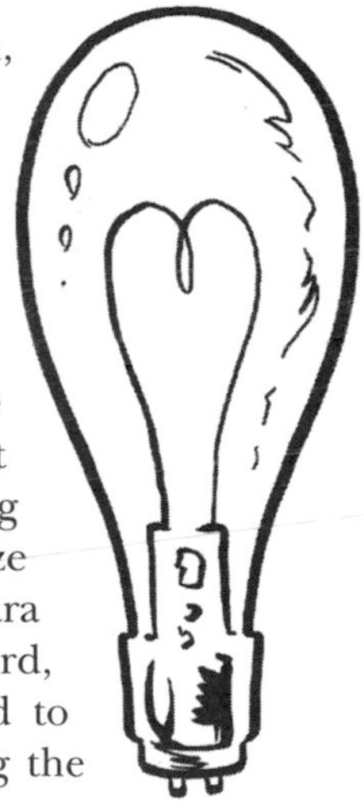

A celebration of American industrialism and innovation, the fair would be lit by electricity. Bids were solicited, and the jousting was nasty—especially between Westinghouse (betting on an AC system) and General Electric (relying on DC). Here was the culmination of the War of the Currents. In the end, Westinghouse underbid all competitors and won the contract.

It was a bold move on many levels. First, Westinghouse stood to lose a fortune. Second, it would be a severe test to prove that Tesla's AC system was capable of lighting what amounted to a whole city. Failure would jeopardize chances of securing a contract for the pending Niagara Falls hydroelectric power generation project. Third, Edison held the patent on light bulbs and threatened to sue Westinghouse should Edison lamps show up among the 250,000 bulbs needed for fairground lighting.

Westinghouse retaliated by inventing his own device—the Westinghouse "stopper" bulb. As opposed to the one-piece fused-glass seal used in the Edison bulb, the Westinghouse innovation employed a ground-glass stopper mated to the bulb envelope for sealing. In the course of just a few months, Westinghouse feverishly produced thousands of these bulbs, a multitude of motors, and an entire two-phase Tesla AC system.

On May 1, 1893, atop a balcony of the fair's Administration Building, President Grover Cleveland ceremoniously pressed a gold-and-ivory telegraph key. A thousand feet away in Machinery Hall, an anxious group of Westinghouse engineers crowded around a 2,000-horsepower steam engine as it roared to life. The massive steam engine powered the Westinghouse generators, which pulsed electricity through the fairgrounds. The crowd roared with delight, as three huge fountains at the Court of Honor sent plumes of water soaring a hundred feet in the air.

For the Westinghouse crew, all was right with the world. The spectacle of buildings, walkways, and fountains gushing with shimmering, color-changing light was a fairyland vision. Westinghouse had succeeded beyond all expectations with Tesla's AC system. It was a resounding defeat for Edison and the death knell of the War of the Currents.

For the time being, Tesla remained behind the scenes and largely unheralded. When the Electricity Building opened on the fairgrounds in June, Westinghouse and Tesla would have the opportunity to beguile the American public with their vision of cheap power. It was a dream almost too momentous to imagine, revolutionizing the way people manage the physical world, how they spend their evening hours, and the very nature of work and leisure. Here for the first time, millions would see the electrical motors that would take over the burdensome physical tasks long performed by man or his animals, and the lamps that would light their houses.

Inside the Electricity Building, inventors rolled out their latest electrical gadgets, many of them dubious: electrically charged belts for a better sex life, electric body invigorators, and electrical hairbrushes. Edison displayed his phonograph and dazzled visitors with an eight-foot, half-ton incandescent light bulb. Ironically this was one of few Edison bulbs on display at the fair, since Westinghouse's stopper lamps were illuminating the grounds.

Tesla, too, would have his turn in the limelight. Soon he would take center stage as America's foremost Wizard of Physics. For the time being, he was back at his New York lab continuing his high-oscillation research. It would take a visit by Westinghouse to convince him how important it was to demonstrate his AC polyphase motor and machinery before the International Electrical Congress. Westinghouse reasoned that by demonstrating the AC systems, its commercial motors, and his oscillators, Tesla would cement his credibility with the foremost electricians in the world.

Tesla took the bait. He journeyed to Chicago that summer where a large part of the Westinghouse exhibit in the Electricity Building was devoted to his machinery. The display featured wireless-illuminated glass tubes shaped into the names of distinguished physicists; high-frequency currents sparkled magically around the room. According to the account of Thomas Commerford Martin, the apparatus represented the fruits of Tesla's labor over ten years:

> *It embraced a large number of different alternating motors and his earlier high frequency apparatus, disruptive discharge coils, and high frequency transformers. Among them a large ring intended to exhibit the phenomena of the rotating magnetic field [a large Egg of Columbus]. The field produced was very powerful and exhibited striking effects, revolving copper balls and eggs and bodies of various shapes at considerable distances and at great speeds.* (The Inventions, Researches, and Writings of Nikola Tesla)

Although many of the devices appealed specifically to the thousand electrical cognoscenti in attendance, Tesla was about to dazzle the assembled throng with a display of unparalleled showmanship at his August 25 lecture. He looked gaunt and exhausted as he took the stage in a beautifully tailored gray-brown four-button cutaway suit. To anyone who noticed, his shoes were unusual—thick-soled with what looked like cork. Gauging the crowd like a seasoned showman, he delivered his lecture on mechanical and electrical oscillations at a methodical pace, supporting his theories with oscillators that could be used to transmit information or electrical energy. Building suspense, he beguiled the crowd with an Egg of Columbus that simultaneously spun balls, pivoted discs, and animated other devices that demonstrated principles of the rotating magnetic field and his theory of planetary motion. And then he put on a show, flashing giant sparks, lighting all sizes and shapes of protofluorescent lamps, and finally lighting himself up with a million volts of electricity passing through his body. Unharmed, he dramatically disproved Edison's charge that alternating current was deadly. It was theater of the highest order (Jill Jonnes, *Empires of Light: Edison, Tesla, Westinghouse, and the Race to Electrify the World*).

Over 25 million people attended the World's Fair, and many were seeing electricity for the first time. The excitement that Tesla ignited at the World's Columbian Exhibition has continued to influence electrical invention to the present day. His spectacle brought waves of adulation and made him a media darling in the burgeoning years of newspaper and magazine publishing. Society at large, people in influential places, scientific intelligentsia worldwide, and, most importantly, potential investors wanted a piece of him.

Elated but exhausted, Tesla returned to New York. He had a lot of work in front of him and needed large amounts of money to fund his research. T.C. Martin stepped in to orchestrate the next step in fund-raising. With the inventor's cooperation, he collected ten years of Tesla's lectures, articles, and discussions and readied them for publication. The compendium detailed all of Tesla's inventions up to December 1893, particularly those bearing on polyphase motors and the effects obtained with currents of high potential and high frequency.

Yet even Martin was reluctant to include material on Tesla's work on wireless transmission. Like many others, Martin considered the prospects far-fetched. As Tesla would recount in a 1919 article for the *Electrical Experimenter,* he was prevented from publishing all but the scantest information on the wireless transmission of energy.

I only need to say that as late as 1893, when I had prepared an elaborate chapter on my wireless system, dwelling on its various instrumentalities and future prospects ... friends of mine emphatically protested against its publication on the ground that such idle and far-fetched speculations would injure me in the opinion of conservative business men. So it came that only a small part of what I had intended to say was embodied in my address of that year before the Franklin Institute and National Electric Light Association under the chapter "On Electrical Resonance." This little salvage from the wreck has earned me the title of "Father of the Wireless" from many well-disposed fellow workers, rather than the invention of scores of appliances which have brought wireless transmission within the reach of every young amateur and which, in a time not distant, will lead to undertakings overshadowing in magnitude and importance all past achievements of the engineer.

Martin's manuscript, running nearly 500 pages, was published in 1894 as *The Inventions, Researches, and Writings of Nikola Tesla.* In the short term, the publication bolstered Tesla'scredibility and spurred new interest among potential investors. Beyond that, the collection documented Tesla's pioneering work for succeeding generations of scientists and inventors.

Chapter 6

ELECTRIFYING DREAMS

If we use fuel to get our power, we are living on our capital and exhausting it rapidly. This method is barbarous and wantonly wasteful and will have to be stopped in the interest of coming generations.... The inevitable conclusion is that water power is by far our most valuable resource. On this humanity must build its hopes for the future. With its full development and a perfect system of wireless transmission of the energy to any distance, man will be able to solve all the problems of material existence. Distance, which is the chief impediment to human progress, will be completely annihilated in thought, word, and action. Humanity will be united, wars will be made impossible, and peace will reign supreme.

—Nikola Tesla

IF HIS LECTURES, EUROPEAN AND CROSS COUNTRY TRAVEL, publications, patent filings, mother's death, nervous breakdown, residential and laboratory relocations, ongoing research, and discoveries in the wireless transmission of energy during the period 1892–1893 were not daunting enough, Tesla was about to enter into a milieu that stretched him even further. It was one thing to concentrate so much of his energy on engineering research and the scientific community. Now he had to face the social elite, the Wall Street financiers, and the countless hangers-on that sought a firsthand audience with him. Often after working all day on a project, he would take dinner at posh Delmonico's restaurant and invite celebrities back to his lab for some startling late-evening demonstration. On one occasion, he brought Mark Twain to tears by relating how the author's works were instrumental in healing Tesla's dire illness as a teenager. The recounting of this incident cemented a lifelong friendship between the two men. Twain relished the opportunity to promote one of Tesla's seemingly risky inventions.

At one evening visit to the lab, according to an oft-repeated story, Twain asked for a demonstration of the benefits of Tesla's work with therapeutic high-frequency oscillators. The famous author ended up furnishing the evening's entertainment when he insisted upon experiencing the gyrations of a platform mounted on an electrified oscillator. Tesla tried to dissuade him, or pretended to, which made Twain all the more determined to prolong the test. Once mounted on the machine, he kept repeating, "More, Tesla more!" Finally, however, he cried out for help, since (as Tesla well knew) one of the effects of such oscillations on the human body was to create turmoil in the bowels.

While working closely with Thomas Commerford Martin on *The Inventions, Researches, and Writings of Nikola Tesla,* the inventor was introduced to Robert Underwood Johnson and his wife Katharine. As editor of The *Century Magazine*, arguably the most popular periodical in the United States at the time, Johnson wielded an extraordinary amount of influence. He and Katharine counted among their circle of friends some of America's best-known artists, writers, politicians, and wealthy industrialists.

The Johnsons were intellectually curious, patronized the arts, and enjoyed stimulating conversation. Tesla was a perfect match for their friendship and a frequent guest at their Lexington Avenue soirées. They became lifelong friends and exchanged frequent letters, including some amorous ones from Katharine to Tesla. Not least among their shared interests was the paranormal, a topic of earnest conversation. Katharine Johnson was taken by Tesla's Old World charm but, since she was married, she tried to pass him off to other rich doyennes—to no avail. Tesla was very much married to his work and avoided romantic entanglements.

Evenings with the Johnsons—to whom he affectionately referred as Luka and Madam Filipov, after a favorite Serbian poem titled *Luka Filipov*—were an opportunity for Tesla to mingle with the social and cultural elite. Among those he met there were authors Rudyard Kipling and Mark Twain; actors Sarah Bernhardt, Eleonora Duse, and Joseph Jefferson; pianist Ignace Paderewski; composer Antonin Dvorak; naturalist John Muir; and architect Stanford White.

Through his associations with Martin and Johnson, Tesla's name appeared frequently in the popular press and his articles gained high exposure. *The Inventions, Researches, and Writings of Nikola Tesla* received strong reviews both stateside and abroad, and went into a number of editions.

In early 1894, during some of his evening forays at the laboratory with Martin and Johnson, Tesla experimented with early photography by phosphorescent light. He posed for several images himself and others with Twain, Jefferson, and Johnson. Several of the photos made their public debut in Martin's article about Tesla in the April 1895 issue of *The Century Magazine.* Tesla would go on to write his own article for *The Century,* titled "The Problem of Increasing Human Energy," in June 1900.

Harnessing Niagara Falls

If Tesla was basking in the limelight, Westinghouse was never one to stand still. As early as fall 1890, he had his eye on a Niagara Falls power project instigated by the Cataract Construction Company. At the start, it was proposed as a hydraulic waterwheel power system. The plan was to divert water from the top of the falls through a system of canals and tunnels to power the factories in the village below. The water would drain back into the river through a long tailrace tunnel so as not to mar the landscape near the falls.

Operating in the background over the next two years, Westinghouse monitored the plan's progress. He knew that his success in Telluride and a resounding victory at the Chicago World's Fair

CATARACT

Cataract is another word for a huge waterfall, as well as what we have come to know as the clouding of a lens inside an eye.

could give him a distinct advantage when it came time to bidding on the Niagara Falls contract. In the back of his mind was the feeling that Tesla's AC system would prove more fruitful than any other proposal.

With the tailrace already underway, the original project organizers, led by New York attorney William Birch Rankine, realized that completion would take an enormous amount of capital—more than they had. Rankine approached financier J.P. Morgan for help, but he was lukewarm to the project because it lacked top-notch leadership. Morgan would invest only if Edward Dean Adams, a prominent Wall Street investment banker (and direct descendant of presidents John Adams and John Quincy Adams) could be brought in to take control of the finances and engineering. Adams was a major stockholder in the Edison Electric Company and happened to live next door to J.P. Morgan in New York.

TAILRACE TUNNEL

A tailrace tunnel, or runoff, serves the same purpose as an above-ground millrace. It is situated below a water wheel, and through it the siphoned water flows back to the river.

Instituted as president of the Cataract Construction Company, Adams was forced to divest his Edison stock in order to avoid a conflict of interest. He then established the International Niagara Commission to attract the best American and European scientific and engineering minds to advise the project. To head the Commission, he secured the services of Britain's Lord Kelvin himself, still a keen proponent of DC. It was a shrewd move on Adams's part to gather these distinguished people and engage them in a contest. Those with the best ideas for harnessing the power of Niagara Falls to transmit energy across great distances would be awarded financial prizes. Among those who balked at the competition was Westinghouse, who commented, "These people are trying to secure $100,000 worth of information by offering prizes, the largest of which is $3,000. When they are ready to do business, we will show them how to do it."

A number of proposals were submitted, and most were rejected out of hand. When the final bidding opened, there were four that

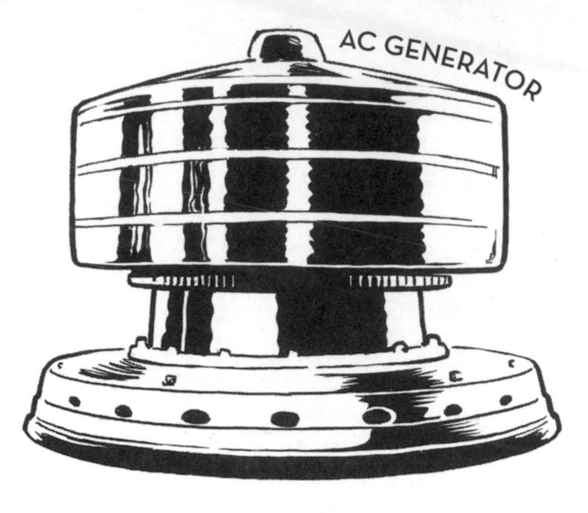

entailed DC power generation and two, from J.P. Morgan's General Electric and from Westinghouse, that would utilize Tesla's polyphase AC system.

Adams and the Niagara Falls Power Company were finally swayed by Tesla's personal appeals and a commitment to transmit power beyond Niagara to Buffalo, and as far as New York City. On May 6, 1893, Adams declared that polyphase alternating current would be their choice for the Niagara project. Six months later, after a series of ugly patent infringement suits between GE and Westinghouse, Adams awarded Westinghouse the contract for building the generators (which Tesla would design) and another contract to GE for building the 22-mile transmission line from Niagara to Buffalo.

In the end, the venture brought investments from some of the wealthiest industrialists in North America in addition to Morgan, such as John Jacob Astor IV, Lord Rothschild, and W.K. Vanderbilt. It was an ambitious and agonizing project, plagued throughout by doubt and financial skittishness. Stanford White was engaged to design what become known as the Edward Dean Adams Power Station to house the great Westinghouse/Tesla generators. From contract to completion, only one person, Nikola Tesla, remained positively convinced that the system would work.

Disaster Strikes

As the Niagara project went forward during the next three years, Tesla continued the exhaustive pace of research at his New York lab, exploring wireless systems to transmit light, power, and communication. William Rankine and Edward Dean Adams were so taken with his ability to keep the Niagara project on course that they offered him $100,000 for a controlling interest in 14 of his patents and the rights to future inventions. Tesla accepted the offer, and the Nikola Tesla Company was founded in February 1895, with Rankin and Adams joining him on the board of directors. The money was much needed, as Tesla had not reaped any funds from the AC system utilized in the Niagara project; he had relinquished all licensing rights to Westinghouse. Fortunately, he still received consultation for his input into the project.

Disaster struck on March 13, 1895, when Tesla's South Fifth Avenue lab was engulfed in fire and the entire building destroyed. His whole world was on the brink of collapse. Equipment, notes, research papers, and years of experiments went up in flames. Although many of his generators and oscillators were housed elsewhere, pioneering work on radio, the wireless transmission of energy, and liquefying oxygen were obliterated. All his thinking would have to be reconstructed from scratch. And he was uninsured to boot.

At first he wandered the streets in despair. An imploring letter from Katharine Johnson, earnest support from Thomas Commerford Martin, and the well-wishes of others eventually lifted Tesla from his despondency. Although he was still bereft of income, financial support from Adams and a burst of self-determination had him searching for a new lab.

Ironically, the first offer of space came from Edison's laboratory in Llewellyn Park, New Jersey, which granted him lab privileges. In late March, he was able to rent space at 46-48 East Houston Street in Lower Manhattan (currently between Lafayette and Mulberry Streets) to house his own lab. Tesla biographer Margaret Cheney would later relate his feelings:

> *I was so blue and discouraged in those days that I don't believe I could have borne up but for the regular electric treatment which I administered to myself. You see, electricity puts into the tired body just what it most needs – life force, nerve force. It's a great doctor, I can tell you, perhaps the greatest of all doctors.*

By July he had the lab up and running and was ready to continue his experiments in the wireless transmission of power and in two new areas, X-rays and radio control. He worked on improving his oscillating transformer to power a new vacuum tube lamp, which he claimed gave out more light and was more efficient than Edison's incandescent lamps. To demonstrate the power of his new bulb, he posed for a famous photo reading James Clerk Maxwell's *Scientific Papers* while seated in a chair given to him by Edward Dean Adams; in the background is a large spiral coil that Tesla was using in his wireless power experiments.

Success at Niagara Falls

While invigorated by his new lab and the work he was doing there, Tesla was mentally anguished by the South Fifth Avenue lab fire. He knew he needed a break from his routine and felt that relief might come from a visit to Niagara Falls. After rejecting invitations for years, he finally made his first visit there in July 1896.

With Westinghouse, Rankine, Adams, other members of the Cataract Construction Company, and a throng of reporters, Tesla toured the power system, which was already up and running. The big question at the time was whether the system would be capable of transmitting electricity to Buffalo. Tesla was upbeat and positive, but being around large machinery and noise made him fidgety with anxiety. He was eager to return to his New York laboratory.

At midnight on November 16, 1896, the switch was thrown that delivered the first power to reach Buffalo, some 22 miles away. Tesla was not there to witness the moment of success, which was a dream come true. As a boy 30 years earlier, he had predicted that he would utilize Niagara Falls for the transmission of power.

The success brought Tesla resounding fame but little profit. Practically every major newspaper and magazine in the country lauded Tesla's achievement as one of the most outstanding engineering accomplishments in history. On January 12, 1897, he returned to Niagara and was sumptuously feted at a banquet in his honor by hundreds of dignitaries, luminaries, and investors in the project. Electricity would soon be lighting up homes, factories, and businesses large and small, as well as powering machinery, rail, and subway systems in Chicago and New York.

NIAGARA FALLS POWER COMPANY

In due course, Lord Kelvin, who had been a strong proponent of DC for the Niagara project, would concede the advantages of AC for long-distance distribution. "Tesla has contributed more to electrical science than any man up to his time," he declared.

If the Chicago World's Fair had marked the end of the War of the Currents, the harnessing of Niagara Falls sealed Tesla's victory.

Dreaming Big

As he accrued more publicity, Tesla's social circle widened even further and with it more women clamored to meet this tall, striking gentleman with the Continental airs, exotic accent, and dapper attire. In today's vernacular, you could say he knew how to work a room, flirt with the ladies, and always spark a conversation. Was it all a mask? Was he playing the bon vivant to attract investors? Women?

By all accounts, Tesla's true comfort zone was the laboratory. His devotion to research and invention stood well above female entanglements. Financial support was most critical to his efforts. While everyone from Westinghouse to the investors in the Niagara project were basking in their success and calculating the wealth they would accrue from their investment, Tesla was, as ever, struggling to finance his next great idea. He was at a crossroads. How could he find backers to develop a wireless transmission system when some of the richest men in America had just poured money into a wired transmission system?

Notwithstanding the consultations on the Niagara Falls project and the disastrous fire at his lab, Tesla had been following the path of his dreams. Much of this was driven by his belief that he could derive energy from the Earth itself. During the intervening years he would do pioneering investigative work on X-rays, earthquakes, Martians, radio, robots, and the wireless transmission of energy.

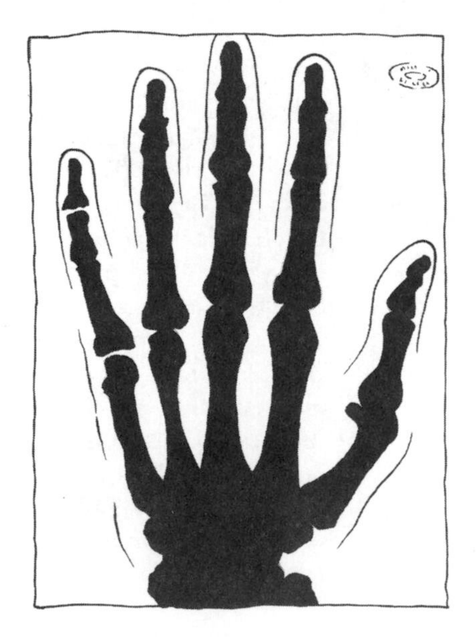

X-Rays

In 1894, Tesla took some long exposure pictures of Mark Twain with the aid of various phosphorescent tubes. Unfortunately Tesla's research in this area was destroyed by the laboratory fire in 1895. That December, when German physicist Wilhelm Conrad Röntgen announced his discovery of X-rays, Tesla remembered that he had salvaged one failed photo from the Twain session. Rather than a picture of the famous author, the image on the glass photographic plate included shadows of the circular lens and the screws on the front of the wooden camera. This image had been captured before the lens cap had been removed for exposure to light.

Upset by Röntgen's announcement, Tesla destroyed the evidence of his work. Gentleman that he was, he granted Röntgen full credit for the invention of the X-ray even though he clearly had produced an earlier example of the process.

Tesla communicated with Röntgen over the course of the next year regarding ways to improve X-rays. He went so far as to self-administer X-rays to his various body parts for potential healing properties, calling the images "shadowgraphs." Although he noticed burns on his skin and abandoned the practice, he maintained that the repeated application of X-rays would have no long-term effects. The damaging effects of radiation were still unknown at the time. As a number of powerful electrical companies began vying for Röntgen's attention, Tesla abandoned his efforts to improve the X-ray process.

Earthquakes

When Tesla sent Mark Twain scurrying for bathroom relief by submitting him to the gyrations of high-frequency oscillators, he had set the stage for future explorations in the transmission of wireless energy. In the years that followed, Tesla returned continually to investigating the phenomenon produced by mechanical oscillators.

In 1898, he reasoned that he could tune his electronic circuits in such a way that the electricity would vibrate in resonance with its circuits—much the way strings vibrate on a musical instrument. Similarly, Tesla visualized mechanical vibrations building up resonance conditions to produce effects of tremendous magnitude on physical objects.

To test his theory, he attached a small mechanical oscillator to an iron pillar on the upper floor of his Houston Street laboratory. As the oscillations were increased, he noticed that certain objects around the room began to vibrate as they came into the same resonance. Tesla described the phenomenon as a "minor earthquake." In fact, the oscillator attached to the iron pillar sent increasing vibrations down through building, into the ground below, and radiating out in different directions with ever-greater force. This was indeed similar to an earthquake, in which the epicenter is calm but the outer reaches experience greater force and destruction.

Unbeknownst to Tesla at the moment of his test, at Police Headquarters on nearby Mulberry Street, chairs began to move of their own accord, objects slid off desks, plaster fell from the ceiling, water pipes burst, and windows shattered. Fearing an earthquake, but used to strange goings-on in the neighborhood, the police dashed to Tesla's lab. Upon their arrival, the courtly inventor was quoted as saying,

> *"Gentlemen, I am sorry, but you are just a trifle too late to witness my experiment. I found it necessary to stop it suddenly and unexpectedly and in an unusual way just as you entered. If you will come around this evening I will have another oscillator attached to this platform and each of you can stand on it. Now you must leave, for I have many things to do. Good day, gentlemen."*

For Tesla, the earth-shaking oscillator test provided a window into his new science of *telegeodynamics.* By utilizing certain vibrations, he theorized, he could literally split the Earth in half. With the correct oscillation, not only could he deal with the transmission of powerful impulses through the Earth to produce effects of large magnitude and distant points, but he could also to detect ore deposits far below the Earth's surface, enemy ships or submarines, or even objects on Mars. Along similar lines, seismologists today are experimenting with time-reversed oscillations to deter potential earthquakes.

Martians

The possibility of life on Mars held many Americans in its grip in the mid-1890s. Among those intrigued by the idea—and eager to find out—was Colonel John Jacob Astor IV, one of the richest men in the world, as well as an investor in the Niagara Company. It was propitious that Tesla shared in this fascination. Astor had presented Tesla a copy of his science-fiction novel, *A Journey in Other Worlds* (1893), which the inventor had enjoyed. Astor himself found time to invent various devices, such as a bicycle brake and an improved turbine engine. Tesla's remarks on signaling Mars might have

seemed outlandish to the press, but they forged a bond with Astor that would serve Tesla in the years ahead. "If there are intelligent inhabitants of Mars or any other planet," he declared, "it seems to me that we can do something to attract their attention.... I have had this scheme under consideration for five or six years." Tesla's reasoning, it will be noted, follows the line of his experiments in mechanical oscillation. As Tesla was quoted in *The Electrical World* (April 4, 1896), sending or receiving signals from Mars

> *... is the extreme application of this principle of the propagation of electric waves. The same principle may he employed with good effect for the transmission of news to all parts of the earth.... Every city on the globe could be on an immense ticker circuit, and a message sent from New York would be in England, Africa and Australia in an instant. What a grand thing it would be in times of war, epidemic, or panic in the money market!*

Radio

Before the fire at South Fifth Avenue, Tesla would set up an oscillating transmitter in the lab and walk around New York with a receiver to see if it would detect signals. Sometimes he received intermittent signals as far north as the Gerlach Hotel on West 27th Street. After the lab was destroyed, while casting about for money and projects, he used some of his credit lines from Westinghouse and Adams to outfit the Houston Street lab and continue wireless research.

Already in his 1893 lecture at the annual meeting of the National Electric Light Association in St. Louis, Tesla had laid out the essential components of a radio system: a transmitter and receiver, antenna, ground connection, and tuning device. That lecture was to be a prelude to more elaborate testing of wireless transmission. The one basic component he had omitted in St. Louis was a speaker, something Tesla would invent and incorporate later. (Among the many physicists of the last quarter of the 19th century who worked on high-frequency electromagnetic vibrations in an effort to invent wireless communication were Germany's Heinrich Rudolf Hertz; England's Sir Oliver Lodge, James Clerk Maxwell, and Sir William Crookes; and the young Italian, Guglielmo Marconi.)

GUGLIELMO MARCONI

In New York around 1897, Tesla would take secret boat trips up the Hudson River with a battery-operated receiver in hand. At distances of over 25 miles, he found, he was able to receive a musical note in tune with the oscillations generated from his Houston Street lab. Ecstatic at the discovery, he drew up his findings and on September 2, 1897, filed for the Patents 649,621 – TRANSMISSION-OF-ENERGY, issued March 15, 1900, and 645,576 – SYSTEM OF TRANSMISSION-OF-ELECTRICAL ENERGY, issued March 20, 1900.

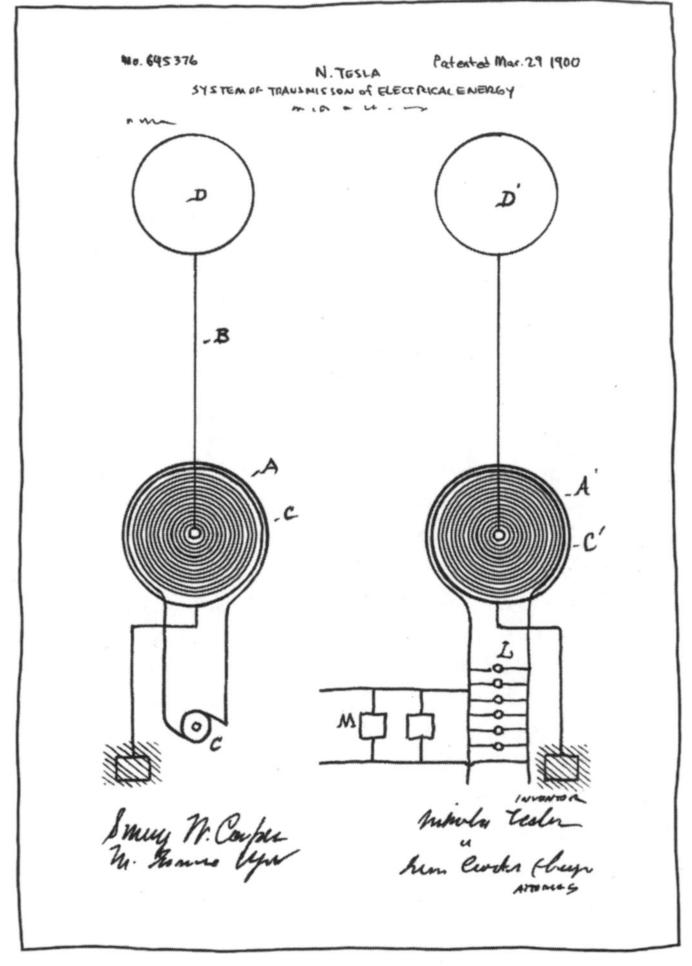

In the application for wireless transmission (649,621), Tesla discusses the need to transmit and receive signals over high altitudes and across unobstructed distances, so that the terminals between transmitter and receiver are "of large surface, formed or maintained by such means as a balloon at an elevation suitable for the purposes of transmission. The practice of tethering balloons to an extreme height might prove impractical." Therefore, he goes on to say,

> *if there be high mountains in the vicinity the terminals should be at a greater height, and generally they should always be at an altitude much greater than that of the highest objects near them. Since by the means described practically any potential that is desired may be produced, the currents through the air strata may be very small, thus reducing the loss in the air.*

It would take another few years for Tesla to lure John Jacob Astor to front him the money to test out these theories in Colorado Springs.

Guglielmo Marconi's first U.S. patent application for a radio, Number 763,772, was filed on November 10, 1900—months after Tesla's. The "Radio Wars" were on! During the next half-century, Marconi would be the major challenger to Tesla in gaining credit for the invention of the radio.

Robots (Teleautomatons)

As always, Tesla was in dire need of money and finally acknowledged that he needed to make a big splash to attract funding. Concurrent with his experiments in wireless radio, he was working on remote-controlled boats. The moment could not have been more perfect to exploit this project. The USS *Maine,* a battleship deployed to protect Cuba and U.S. interests on the island from Spain, mysteriously exploded in Havana Harbor on February 15, 1898. Implicating Spain in the explosion, the United States officially declared war on April 28, 1898. Tesla rushed to file Patent 613,809 – METHOD OF AND APPARATUS FOR CONTROLLING MECHANISM OF MOVING VEHICLE OR VEHICLES on July 1, 1898, stating in his application,

> *I have invented certain new and useful improvements in methods of and apparatus for controlling from a distance the operation of the propelling-engines, the steering apparatus, and other mechanism carried by moving bodies or floating vessels.*

This was his first foray into a field that he would come to call "teleautomation." So novel was the application that the patent office sent an examiner to inspect the boat in action in Tesla's lab. The patent was granted on November 8, 1898.

With patent protection, Tesla was once again ready to amaze the public and lure investors. On December 8, 1898, in an exhibition at New York's Madison Square Garden, he demonstrated his teleautomaton, or radio-controlled boat, to the public. The crowd responded with disbelief. To some witnesses, the boat moved by magic, to others by telepathy. A few suspected a trained monkey was installed in the boat, obeying Tesla's commands. In reality, he used a transmitter box to send signals that shifted electrical contacts in the boat, adjusting the rudder, motor, and lighting.

The demonstration could not have come at a better time, as the Spanish-American War was at its height. Tesla saw an opportunity to sell a radio-controlled torpedo to the U.S. Navy, but the idea was met with derision. Perhaps his mistake was trying to sell the weapon as a means of peace rather than war. He reasoned as follows:

ROBOTS

The distinction between Tesla's boat (a radio-controlled mechanism) and a robot is that a robot is a man-made mechanical device that can move by itself, whose motion is pre-arranged, planned, sensed, actuated, and controlled, and whose moves are influenced by "programming." Tesla's boat, as described in his patent, mimics a robot. It contains a sensing device that responds to instructions from a distant radio.

These automata, controlled within the range of vision of the operator, were the first and rather crude steps in the evolution of the Art of Telautomatics.... The next logical improvement was its application to automatic mechanisms beyond the limits of vision and at great distance from the center of control.

Tesla felt strongly that wars could be waged machine against machine. And that would be progress, he believed, because with automatons there would be no bloodshed. Tesla remained convinced that his idea for remote control would eventually catch on. Unfortunately, he would not live to see the day when his prophecy would become a reality.

The Lap of Luxury

It was around this time that Tesla moved into Astor's recently completed Waldorf Astoria Hotel on Fifth Avenue. It was the epitome of opulence and luxury. Astor treated his friend with the highest esteem, according him such special amenities as his own dining table. But Tesla was broke and living off borrowed money. His research was at a standstill. His New York lab was too small and unsafe for the scale of his projects, and the electricity being generated at the time for the lighting, telegraph, telephone, and transportation systems mushrooming across Manhattan interfered with the wireless transmission experiments he wished to conduct.

Astor, who had been following Tesla's work closely, learned that he required funds to continue his research and invested $100,000 in the Tesla Electric Company. Of this amount, $30,000 went to Tesla directly; the rest went to stock purchases. The injection of money enabled Tesla to take advantage of an offer by Leonard E. Curtis of the Colorado Springs Electric Company to build a temporary plant in that city. This would give Tesla an opportunity to conduct experiments on a gigantic scale. He would be provided with all the land and electrical power he needed for his work. And so, in May 1899, Tesla moved to Colorado Springs.

Chapter 7

FOLLIES ABOUND

Besides machinery for producing vibrations of the required power, we must have delicate means capable of revealing the effects of feeble influences exerted upon the earth. For such purposes, too, I have perfected new methods. By their use we shall likewise be able, among other things, to detect at considerable distance the presence of an iceberg or other object at sea. By their use, also, I have discovered some terrestrial phenomena still unexplained. That we can send a message to a planet is certain, that we can get an answer is probable: man is not the only being in the Infinite gifted with a mind.

—Nikola Tesla

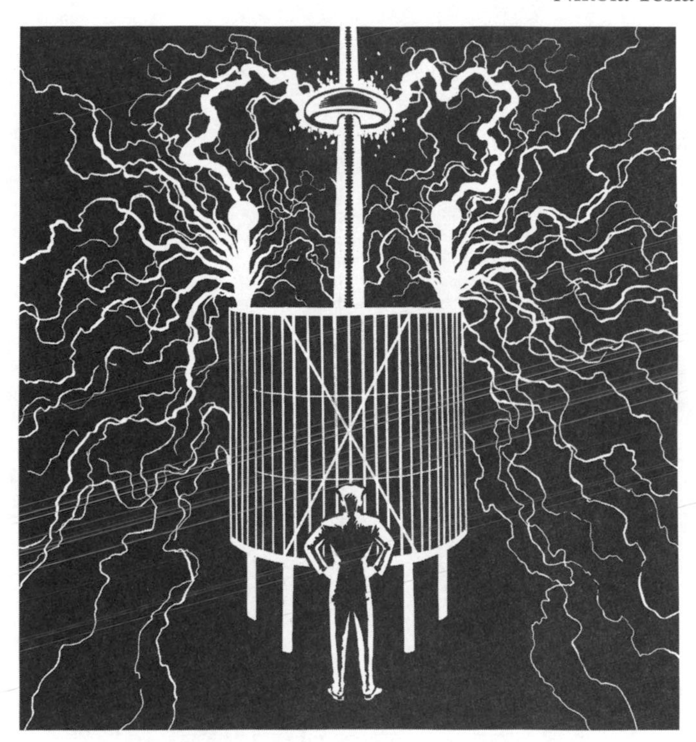

TESLA ARRIVED IN COLORADO SPRINGS on May 18, 1899. Ostensibly, Astor had invested in him to pursue improvements in fluorescent lighting. Instead of adhering to Astor's wishes, Tesla was well into his plans for setting up a massive experiment on the wireless transmission of energy, light, and communication.

With an auspicious welcome from the town's elite, Tesla settled into Room 207 at the Alta Vista Hotel. He promptly instructed the staff to deliver 18 fresh towels to his room each morning. Both the room number and number of linens were in keeping with his obsession that they be divisible by three.

Colorado Springs Experimental Station

Colorado Springs presented Tesla the opportunity to conduct his transmission tests at 6,000 feet above sea level in a dry climate, atmospheric conditions that befitted his experiments. He would be away from the social obligations and prying press of New York. Indeed he would be able to work in secrecy. Upon arriving, he sent a request to his assistant back East, George Scherff to ship electrical equipment and glass blown tubes from the New York laboratory.

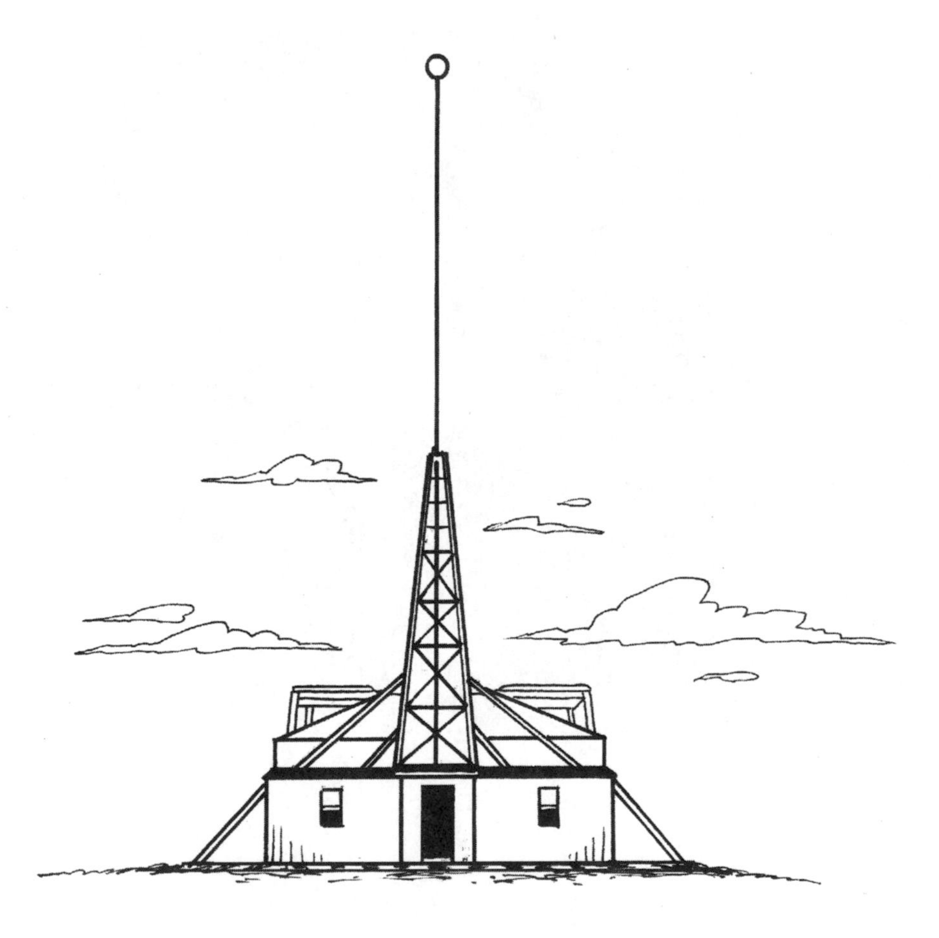

Not only did Leonard E. Curtis provide Tesla with the means to set up the lab; he also secured free electricity for the inventor. The actual plans for the building had been kept secret until only a few days before Tesla arrived in Colorado. Outside town, in a pasture with a view of Pikes Peak, Tesla's odd-looking experimental facility was being constructed under the auspices of local carpenter Joseph Dozier. Tesla and Dozier shared an inclination to discuss paranormal behavior and interplanetary communication.

The main wooden lab measured 50-60 feet by 70-80 feet, and approximately 18 feet high, with a roof that opened to the sky. A smaller space that served as an office jutted out from the front of lab. At first Tesla had planned to fly balloons to lift the wires for sending signals, but the balloons proved incapable of lifting such heavy weight.

As an alternative, he erected a large wooden tower on the roof platform, with a telescopic copper pole through the center. This retractable device, topped by a copper ball 3 feet in diameter, measured nearly 200 feet when fully extended. The pole, which he referred to as his *magnifying transmitter,* rose from the lab's enormous Tesla coil, which he called his *magnifying transformer.* Over time, these two terms came to be used interchangeably to describe the entire system.

Tesla's lab looked extremely out of place on the prairie. Fencing was erected with signs that admonished, "KEEP OUT – GREAT DANGER." This was no idle threat, as Tesla would soon be generating enough electrical power to kill an errant trespasser instantly.

Inside the main laboratory, all manner of transformers and condensers were set up upon their arrival from New York. Together with assistant Fritz Lowenstein, who also came out from New York, Tesla assembled what was then the world's largest Tesla coil. The coil was basically a 6-foot outside circular wall that was approximately 50 feet in diameter. Straddling the top of the wall were a couple of interwoven thick cables, or windings. Taken together, this served as the primary winding for his transformer. Within this ring were a number of secondary coils, wound with differing amounts of turns. At any given time, one end of a secondary coil could be sent into the earth and the other end could be attached to the transmitting pole. In this fashion, Tesla could send different wavelengths to his various receivers. By the time construction of the Tesla coil was completed in mid-June, he could begin experimenting.

Missions Accomplished

Tesla wrote in his Colorado Springs diary that he had come to accomplish three main goals:

- Develop a transmitter of great power;
- Perfect a means for individualizing and isolating the energy transmitted.
- Ascertain the laws of propagation of currents through the Earth and the atmosphere.

Through months of experimentation, he made a number of discoveries that led to the realization of these goals. He was able to discover the Earth's electrical properties, send messages through the ionosphere, and transmit energy great distances in three manners:

1) *The Earth's Electrical Properties—Stationary Waves*

With his magnifying transmitter, Tesla could conduct experiments in wireless communication and energy to any point on the globe. He would begin by measuring lightning discharges during storms and set up oscillations from his coil that were in harmony with the sparks of lightning. By varying his oscillations, he could produce varying wavelengths that could be transmitted and received at different frequencies. His apparatus thus led to the discovery of "stationary" waves in July 1899, which he explained as follows:

When we raise the voice and hear an echo in reply, we know that the sound of the voice must have reached a distant wall, or boundary, and must have been reflected from the same. Exactly as the sound, so an electrical wave is reflected, and the same evidence which is afforded by an echo is offered by an electrical phenomenon known as a "stationary' wave – that is, a wave with fixed nodal and ventral regions [i.e. a sine wave]. Instead of sending sound-vibrations toward a distant wall, I have sent electrical vibrations toward the remote boundaries of the earth, and instead of the wall the earth has replied. In place of an echo I have obtained a stationary electrical wave, a wave reflected from afar. ("The Problem of Increasing Human Energy")

2) Sending Messages Through the Ionosphere

The discovery of stationary waves and utilizing the conductive properties of the Earth meant that Tesla would be able to send messages much farther than his competitor Marconi could. He was tapping into the Earth's geomagnetic pulse. The ionosphere would become Tesla's electrical highway for carrying energy from a transmitting tower to a receiving plant.

In the process of make that discovery, Tesla's highly sensitive equipment also began picking up rhythmic beeps. In a 1901 article for *Collier's Weekly* titled "Talking with Planets," he would write:

> *The changes I noted were taking place periodically, and with such a clear suggestion of number and order that they were not traceable to any cause then known to me. I was familiar, of course, with such electrical disturbances as are produced by the sun, Aurora Borealis and earth currents, and I was as sure as I could be of any fact that these variations were due to none of these causes.... It was some time afterward when the thought flashed upon my mind that the disturbances I had observed might be due to an intelligent control. Although I could not decipher their meaning, it was impossible for me to think of them as having been entirely accidental. The feeling is constantly growing on me that I had been the first to hear the greeting of one planet to another.... I have never ceased to think of those experiences and of the observations made in Colorado. I am constantly endeavoring to improve and perfect my apparatus, and just as soon as practicable I shall again take up the thread of my investigations at the point where I have been forced to lay it down for a time.* ("Talking With the Planets")

In 1955, rhythmic one-two-three beeps were discovered to be emanating from the planet Jupiter. Perhaps these were the sounds that Tesla had attributed to signals from Mars, as suggested by research in the 1990s by electrical engineer James F. Corum and his brother, physicist Kenneth L. Corum.

3) *Transmitting Energy*

Having paved the way for wireless communication, Tesla embarked on experiments to transmit electricity wirelessly. He would take his receivers a considerable distance from the lab, with Lowenstein left behind to throw the switch on the transmitter at varying intervals. Tesla would report back when the receivers detected energy. However, since Lowenstein was back in the lab, unable to witness what Tesla reported, there is no actual confirmation of these experiments.

Sometime in early fall 1899, Lowenstein left for Europe. Tesla then summoned another assistant from New York, Kolman Czito, who would remain with him well into old age for both men. Tesla would prepare Czito for some of his boldest experiments.

Wanting visual proof that his experiments were proving fruitful, Tesla contacted Robert Underwood Johnson to write an article for

The Century, with photos to demonstrate his achievements in Colorado Springs. In late December 1899, photographer Dickenson Alley arrived from New York to shoot an astonishing set of photographs. One iconic image was achieved through the technique of multiple exposures. The photo depicts a seated Tesla reading a book amid hundreds of streamers of electricity streaking around him. If the photograph were shot in real time, Tesla would have been electrocuted on the spot.

For the boldest of experiments, Tesla and Czito planted numerous incandescent lamps in the fields beyond the laboratory. Upon Tesla's command, Czito threw the transformer switches for a brief moment while Tesla stood in the lab doorway, observing huge bolts of electricity issuing from the top of the transmitting tower. The sound of millions of volts of streaming electricity was thunderous—then everything went silent. Tesla's experiment had short-circuited the power at the Colorado Springs Electric Company's generating station, plunging the surrounding community into complete darkness.

The disaster contributed to Tesla's departure from Colorado Springs. He was once again in debt, with a stack of outstanding bills. The Johnsons entreated him to return to New York for the Christmas holidays. Tesla was anxious to do so but was not be able to depart until January 7, 1900.

Back in New York

Ironically, when Tesla arrived back in New York, Marconi was also there soliciting funds for his wireless system. Marconi's confidant was the Serbian physicist Michael Pupin, who had worked with Edison in challenging Tesla's AC patent. Now, in early 1900, Pupin succeeded in bringing Tesla and Marconi together for an uncomfortable meeting at Tesla's lab to pick each other's brains. Tesla was confident that he possessed the ability to transmit messages around the globe. Marconi didn't believe him. The upshot of the meeting was that Pupin gained an understanding of Tesla's theories and pirated the principles in order to gain employment with Marconi. Pupin and Marconi's machinations would play out negatively against Tesla in the years ahead.

Tesla was less concerned with Marconi's current experiments than with the need to secure investments for a massive broadcast and power transmitter that would capitalize on his Colorado Springs experiments. He faced the same conundrum as he had with his invention of the AC induction motor. Without the support system and acumen of previous patronage, he would have to go it alone—devise his own strategy for filing patents, promote his wireless transmission system, and garner interested investors. Years earlier it was sufficient to produce working motors that introduced his AC ideas to the world. Now in 1900 his theories were on a global scale. The stakes were much higher, and the dollar amount would have to be much higher as well.

In order to protect his Colorado Springs findings, Tesla set about drawing up and filing a set of four important patents between mid-

May and mid-July, 1900. The first one was easily the most brilliant, in which Tesla set forth his discovery and utilization of stationary waves to transmit power and send messages through the earth (Patent 787,412 – ART OF TRANSMITTING ELECTRICAL ENERGY THROUGH THE NATURAL MEDIUM).

Tesla's efforts to interest the U.S. Navy in the teleautomation torpedo and wireless communication systems met with bureaucratic rejection. His repeated attempts to obtain new investments from John Jacob Astor were greeted by a wall of stony silence. George Westinghouse was uninterested in wireless projects but did extend Tesla a few thousand dollars in credit. Even Tesla's appeals to the New York elite at his usual haunts—the Player's Club, Delmonico's, and the like—failed to rouse backers.

"The Problem of Increasing Human Energy"

With no practical inventions to show for his Colorado Springs experiments, Tesla was hard-pressed to find financial backers for future development in wireless transmission. The lack of anything practical to demonstrate was problematic. The years between Tesla's initial AC-driven motors and the turn of the century gave rise to a host of well-trained electrical engineers and a myriad of inventors who profited greatly from Tesla's initial discoveries. Less apparent to investors in 1900 were the benefits to be gained from a system that would transmit energy wirelessly.

While filing his wireless patents, therefore, it was important for Tesla to decide how to promote his theories in order to attract

THE PROBLEM OF INCREASING HUMAN ENERGY

potential investors. Thomas Commerford Martin, who had been instrumental in arranging earlier presentations to the electrical engineering community, no longer could be relied on, as he was skeptical of Tesla's grand predictions. The other alternative was for Tesla to appeal directly to the public in the hope of luring investors. Gambling on the latter approach, he wrote a long article in florid language that described the global implications of his magnifying transmitter. It was designed, he declared, to solve all the world's energy needs. After much give and take, Tesla convinced Robert Underwood Johnson to publish his treatise, "The Problem of Increasing Human Energy," in the June 1900 issue of *Century Magazine.* Running more than 70 pages, the article was accompanied by startling photographs of Colorado Springs and Tesla's automaton.

J.P. Morgan's $$$$$$

Tesla's treatise in *Century Magazine,* interwoven with his ramblings on lasting peace, harnessing the sun's energy, and interplanetary communication, proved highly provocative. Members of the public greeted his idea of a World Wireless System as visionary, but the scientific community was less than enthusiastic. The lengthy

JOHN PIERPONT MORGAN

essay failed to attract investors, though it might have piqued the interest of the richest man in America, J. Pierpont Morgan. Among the ideas that intrigued him was a magnifying transmitter capable of locating ships at sea and transmitting their position back to the mainland. The prospect appealed to Morgan, who was a yachtsman. Indeed his own negotiations with Marconi for a secure system had failed.

Tesla and Morgan both had close social ties with the Robert and Katharine Johnson, at whose social gatherings the inventor and industrialist appeared together over the ensuing months. Morgan, who had invested in the Niagara project, was well aware of Tesla's success with it. No doubt Tesla got Morgan's ear about the telegraphy system he proposed that would broadcast securely over multiple wavelengths, compared with Marconi's single-wavelength system. The series of discussions and entreating letters from Tesla finally moved the tycoon to hand Tesla a check for $150,000. In return, Morgan received a 51% share of Tesla's lighting and wireless patent rights.

Morgan's $150,000 was little more than pocket change for an industrialist whose net worth in the steel industry alone ran into a billion dollars. Tesla was excited that he could move ahead with his plans to build a wireless telegraphy system, but perhaps he misconstrued the arrangement with Morgan as the beginning of a bigger, longer-term partnership. With the thousands of dollars it took to light the Chicago Exposition and the millions of dollars it took to complete the Niagara project, Tesla could hardly have expected to accomplish his goal of a wireless telegraphy system with the initial funding from Morgan. His grandiose plans for a completely wireless world broadcasting transmission system would require much more.

Wardenclyffe

Morgan's commitment enabled Tesla to negotiate with banker and lawyer James S. Warden for a 200-acre tract of Warden's land on Long Island, New York. Portraying the yet-to-be-built transmission system as the centerpiece of Warden's utopian real-estate enterprise, Tesla convinced him to deed over the parcel of land. Today this part of Long Island is known as East Shoreham; back then the projected development was named Wardenclyffe.

Tesla was able to enlist the services of his architect friend Stanford White, the designer of the power plant at Niagara Falls, to draw up architectural plans for a two-story laboratory and powerhouse that still stands on the site today. The tower itself was designed and constructed by White's associate, W.D. Crowe of East Orange, New Jersey. Construction began in September 1901. Tesla would describe the tower as

> *187 feet high, having a spherical terminal about 68 feet in diameter. These dimensions were adequate for the transmission of virtually any amounts of energy.... The transmitter was to emit a wave-complex of special characteristics and I had devised a unique method of telephonic control of any amount of energy.* (My Inventions)

Theoretically, the transmission tower would send signals wirelessly through the Earth, which acts as a conducting body, to a receiver anywhere around the globe.

During construction, Tesla would commute by railroad each day from his residence at the Waldorf Astoria, along with a servant who carried a lavish lunch prepared by the hotel's staff. They would arrive at 11:00 A.M. and depart at 3:30 P.M. after Tesla's supervision of the day's construction.

As the laboratory neared completion and equipment arrived from the Houston Street lab, Tesla took up residence in a nearby bungalow for a year. Outfitting the lab was particularly stressful for him, with delays in the manufacture and shipment of machinery and irregular intervals in the receipt of money from Morgan.

Erecting the Tower

The proposed tower would be located 350 feet from the laboratory in order to protect its occupants from errant streams of lightning emanating from the mushroom-shaped top. The new tower posed a number of design and engineering problems. First, no wooden structures of this height that employed minimal metal had ever been built before. The top of the tower was so large that it would act as a sail in stiff winds and easily topple the entire structure. When finally completed in 1902, the tower loomed high over the Wardenclyffe facility like a bizarre erector set topped by a huge mushroom.

In order for Tesla to achieve the objective of transmitting electrical power through the Earth, he reasoned that he would need a ground connection from the transmitting tower. To this end, he sank a 10-foot by twelve-foot well 120 feet under the tower and below the water table. Then, to complete the ground, he connected a metal shaft from the mushroom cap of the tower to the bottom of the well. An elaborate system was also necessary to properly connect the primary coils and the magnifying transmitter's secondary coil. Patent 1,119,732 – APPARATUS FOR TRANSMITTING ELECTRICAL ENERGY provided an illustration and principles for Wardenclyffe Tower.

The big question was would it work?

Complicating matters, Marconi claimed to have sent the three-dot, Morse code signal "SSS" across the Atlantic on December 12–13, 1901, heralded in The *New York Times.* Thomas Commerford Martin switched his allegiance from Tesla to Marconi, arranging for the latter to address the AIEE about his accomplishment. Tesla brushed off the announcement with a glib rejoinder: Marconi was merely sending signals, whereas he would be able to transmit electrical power wirelessly.

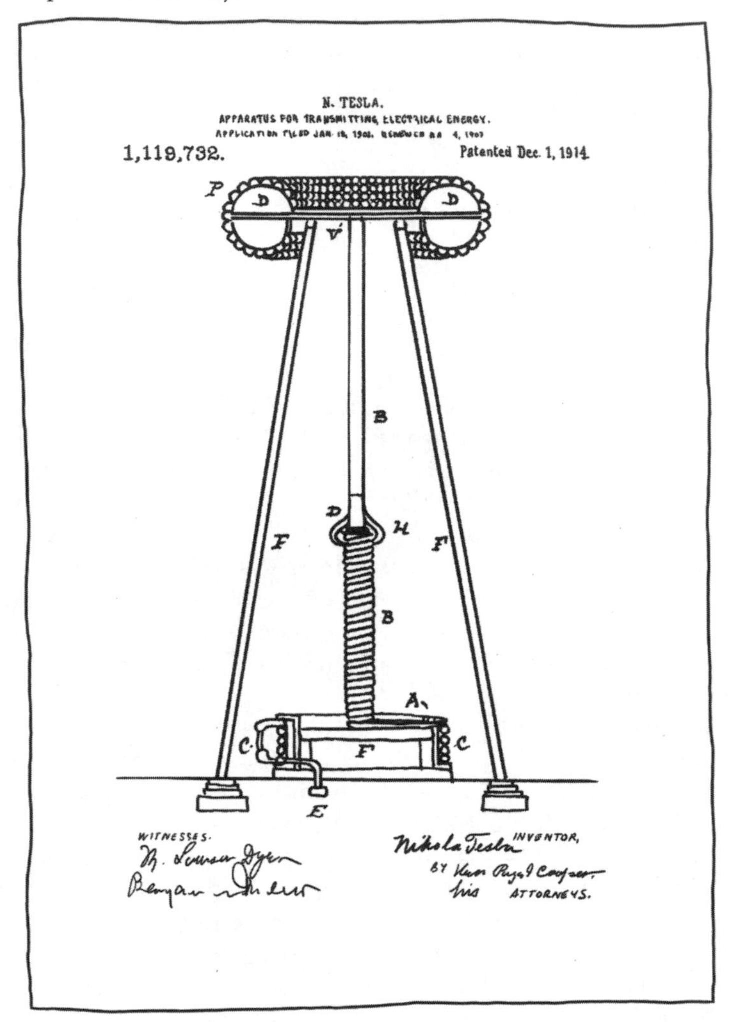

Morgan, also sensing that Marconi was ahead of Tesla in creating a wireless telegraphy system, felt little motivation to provide Tesla with additional funding. Moreover, Tesla had not lived up to his contract with Morgan to deliver a transatlantic telegraphy system. Indeed, Tesla had hoodwinked Morgan into believing that he was building the communication system while in fact he was working on his wireless energy dream.

Nor was Marconi the only inventor whose publicity campaign pushed him ahead of Tesla. Other notable entrants in the field included Elihu Thomson, Michael Pupin, and rival wireless inventor Lee de Forest, who made major innovations in broadcasting and sound-on-film development. All of their advances were fundamentally inspired by Tesla's discoveries.

In 1902, Tesla stepped up his entreaties to Morgan to pour more money into the Wardenclyffe enterprise. In a series of earnest letters to the financier, Tesla appealed to Morgan's vanity and business instincts. Tesla outlined a "World Telegraphy System" for which Morgan could potentially reap great profits from all the receivers that would have to be built around the world.

> *It makes possible not only the instantaneous and precise wireless transmission of any kind of signals, messages or characters, to all parts of the world, but also the inter-connection of the existing telegraph, telephone, and other signal stations without any change in their present equipment. By its means, for instance, a telephone subscriber here may call up and talk to any other subscriber on the Globe. An inexpensive receiver, not bigger than a watch, will enable him to listen anywhere, on land or sea, to a speech delivered or music played in some other place, however distant.* (My Inventions)

Was this just grandiose boasting?

No matter what, Tesla drove himself to complete as much of the tower as possible. He had to begin testing in order to prove the viability of his concept. *The New York Sun* reported on July 16, 1903:

> *Natives hereabouts are intensely interested in the nightly electrical display shown from the tall tower where Nikola is conducting his experiments in wireless telegraphy and telephony. For a time, the air was filled with blinding streaks of electricity traveling through the darkness on some mysterious errand.*

Morgan, for his part, was too deeply involved in his own myriad interests in the shipping, steel, and railroad industries to pay much attention to Tesla. The inventor's incessant pleas for money eventually became full of vituperation and anger toward the industrial baron. Morgan flatly refused to become involved in any more of Tesla's schemes; the great enterprise at Wardenclyffe still had not demonstrated the ability to turn a profit. In all probability, the many speculators then entering the broadcast field also contributed to Tesla's diminished creditability. Although Morgan now viewed him as a poor risk, Tesla still believed that

> *the true rewards are ever in proportion to the labor and sacrifices made.... I feel certain that of all my inventions, the Magnifying Transmitter will prove most important and valuable to future generations. I am promoted to this prediction not so much by thoughts of commercial and industrial revolution which it will surely bring about, but of the humanitarian consequences of the many achievements it makes possible.* (My Inventions)

Broke and Broken

Without Morgan's capital and endorsement, Tesla went into a downward spiral. As much as he tried to keep working, the bills mounted and his creditors besieged him. His debts in Colorado Springs had not been fulfilled. He could not pay Westinghouse for the dynamos and other machinery loaned to him on credit. His current employees were going without pay and began to leave. Before long, only his loyal assistant George Scherff remained to help with the research. Even that came to a halt when Tesla could no longer pay for the coal to fuel the boilers.

Financially overwhelmed and psychologically overcome, Tesla was forced to abandon Wardenclyffe in late summer 1905. He apparently suffered a complete nervous breakdown, clinging to his vision for world wireless energy. He returned to New York City to live on credit at the Waldorf Astoria.

Tesla's personal world underwent further transformation in the succeeding years with the passing of some of the important figures in his life. Stanford White, who had designed both the Niagara Falls and Wardenclyffe powerhouses, was famously murdered by Harry K. Thaw over a lover's dispute with showgirl Evelyn Nesbitt in 1906. Mark Twain would succumb in 1910. John Jacob Astor was lost in the sinking of the *Titanic* in 1912. George Westinghouse passed on two years later.

Besieged by creditors, Tesla resorted to desperate measures. To guarantee payment of his hotel bills, he gave two mortgages on Wardenclyffe to the proprietor of the Waldorf Astoria, George C. Boldt. Unable to make any payments at all, Tesla turned over the full deed in 1915.

Wardenclyffe would be Tesla's last undertaking of such magnitude. While the tower had a futuristic science-fiction aspect, he left no clear plans or description of the full operating system.

Chapter 8

RIDING THE ROLLER COASTER: NEW INVENTIONS, BANKRUPTCY, HONORS

You propose to honor me with a medal which I could pin upon my coat and strut for a vain hour before the members and guests of your [AIEE] Institute.... And when you would go through the vacuous pantomime of honoring Tesla you would not be honoring Tesla but Edison who has previously shared unearned glory from every previous recipient of this [Edison] medal.

—Nikola Tesla

THE DEMISE OF THE WARDENCLYFFE PROJECT was unbearable for Tesla. Dodging creditors, getting by on small loans from friends and assistant George Scherff, and the inability to pursue the tangle of patent infringements from which others benefited all left Tesla bereft. Scherff coaxed him repeatedly to develop some minor inventions that could be marketed for profit. Tesla, still obsessed with improving the world on a grand scale, could not be bothered with such trifles. He continued to seek funding, but the Panic of 1907 sent potential investors in other directions. The moneyed elite were more interested in saving their own institutions.

Persevering

Seeking relief from his mental anguish, Tesla would secretly journey out to the abandoned Wardenclyffe site. There, on the brink of tears, he would self-administer electric shock therapy.

Although the waning days of Wardenclyffe signaled the death knell to Tesla's World Telegraphy System, it did not leave him with a dearth of new ideas. He finally scraped together rent money for a tiny office in the Singer Sewing Machine Building at 165 Broadway (today the site of 1 Liberty Plaza). To generate a modest income, he wrote a series of visionary articles with a tone that was alternately passionate and embittered. For wasn't he the great discoverer of AC and the inventor of the AC induction motor, fluorescent lamps, remote control, wireless telegraphy, and a host of other electronic improvements? Shouldn't investors flock to underwrite his ideas?

With no benefactors forthcoming, he realized he needed a different strategy.

Turbines

During the dark days of his recovery, Tesla's interests reverted back to his childhood memories of flying. While the Wright Brothers had initiated the era of human-powered flight in 1903 with a lightweight gas engine, Tesla believed that airplane engines had to be lighter, smaller, and subject to less friction. He also sought to eliminate propellers, which themselves increased resistance to propulsion.

Borrowing again from the thinking that led to his invention of the AC induction motor, Tesla would return to another childhood memory—that of a spinning waterwheel in a flowing river. Just as the back-and-forth rotating magnetic field played a prominent role in his AC motor, he realized, so the flow of fluid could proceed in his bladeless turbine.

Technically we think of a turbine engine as essentially a shaft with fanlike blades. Applying pressure to the blades causes the shaft to turn at a high speed, generating power. To Tesla's mind, however, the technology is inefficient because the blades themselves create a resistant force. To overcome this additional resistance, Tesla explored the principles of viscosity and adhesion. He eliminated the blades and arranged a series of closely packed parallel metal disks mounted on a shaft within a sealed chamber. When water or compressed air entered the chamber and passed between the disks, the disks turned. That rotated the shaft. Because the fluids have viscosity and all fluids molecularly adhere to solids, the movement of the water along the surfaces of the disks caused the disks to rotate. If the fluid entered at the center of the turbine and goes out the periphery, the turbine acted as a pump. If the fluid entered at the periphery and exited out at the center, the turbine acted as a motor. Thus, the turbine could be an efficient engine or a switch for powering a pump or compressors, for running automobiles and aircraft.

Working with Julius Czito, the son of his former assistant at Colorado Springs, Tesla produced their first turbine prototype in 1906.The motor weighed less than 50 pounds and was able to deliver 110 horsepower. Tesla would describe this small but highly efficient turbine as a "powerhouse in a hat." The workings of the invention are described in Tesla Patent 1,061,206 – TURBINE, filed in 1909. In an interview with *The Washington Post*, Tesla called the bladeless turbine "the greatest of my inventions.... Here you have a new power for pumps, steam engines, gasoline motors, for automobiles, for airships, for many other uses, and all so simple."

One problem with the design was that the metallurgic materials available to him at the time could not withstand the rigors of the fluids applied to them and thus would distort the parallel disks. Thus, if he were to produce a turbine that measured up to his exacting standards, he would need money for further research. Fortunes had already been invested in existing turbines, however, so there was little impetus for capitalists to change the status quo. And so, once again, Tesla had developed an invention that would have far-reaching implications for industry and the benefit of mankind, but once again he faced financial obstacles and derision.

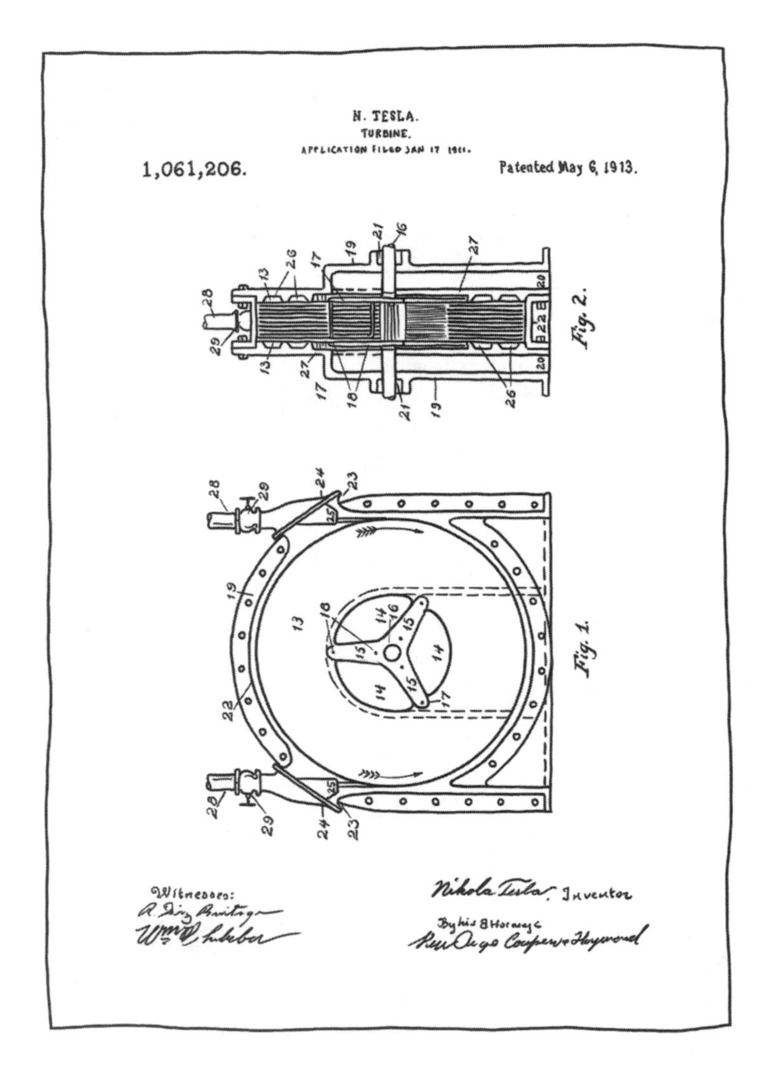

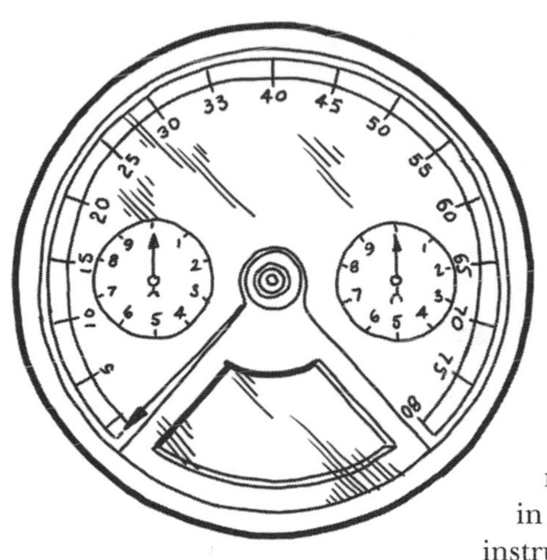

Speedometers

Although the turbine proved unsuccessful as a motor in his lifetime, Tesla was able to attract some manufacturing interest in the use of his invention as a pump. In particular, the Waltham Watch Company saw his research as potentially beneficial in the manufacture of precision instruments. Waltham's seed money led directly to Tesla's invention of the speedometer, whose key principle he succinctly described in Patent 1,274,816 – SPEED-INDICATOR:

> *I have described a new type of speed measuring instrument wherein the adhesion and viscosity of a gaseous medium, preferably air, is utilized for torque-transmission from a primary driving to a secondary pivoted and torsionally restrained member under conditions such that the rotary effort exerted upon the latter is linearly proportional to the rate of rotation of the former. The principles of that invention find place in my present construction [the speedometer].*

Unfortunately for Tesla, the patent granted in 1918 was assigned to the Waltham Watch Company. With his customary lack of business sense, Tesla came away with nothing after years of research and development. Today, Tesla's bladeless corrosion-defying pump is widely used as a means of pumping fluids in oil fields and retrieving viscous materials that contain solids other pumps can't handle.

Flying Stoves, Flivvers, and VTOLs

With renewed confidence if not financial stability, Tesla rented office space in the Metropolitan Life Tower on Madison Square.

FLIVVER

Early 20th-century slang term for a cheap, mass-produced automobile; it could also be applied to a similarly crude, rough-riding airplane of the day.

Boldly predicting that jet planes would be the aircraft of the future and that his bladeless turbine would power them, Tesla courted John Jacob Astor because of his interest in flight. Astor saw through the ruse. Behind Tesla's appeal lurked the inventor's intention to develop wireless energy and propel an electric turbine with power supplied from stations around the world. Astor refused investment capital, as did George Westinghouse.

Given his propensity for hyperbole, Tesla told one Westinghouse employee, "You should not be at all surprised if some day you see me fly from New York to Colorado Springs in a contrivance which will resemble a gas stove and weigh as much." Tesla's flying machine was to be an improvement of Henry Ford's Model-T of the air called the "flying flivver."

Even without funding from Astor or Westinghouse, Tesla for years continued his theoretical research into a flying machine that could be powered by his bladeless turbine. In Patent 1,655,114 – APPARATUS FOR AERIAL TRANSPORTATION, filed October 7, 1927, he detailed his proposal:

> *The invention consists of a new type of flying machine, designated "helicopter-plane", which may be raised and lowered vertically and driven horizontally by the same propelling devices and comprises: a prime mover of improved design and an airscrew, both especially adapted for the purpose, means for tilting the machine in the air, arrangements for controlling its operation in any position, a novel landing gear and other constructive details, all of which will be hereinafter fully described.*

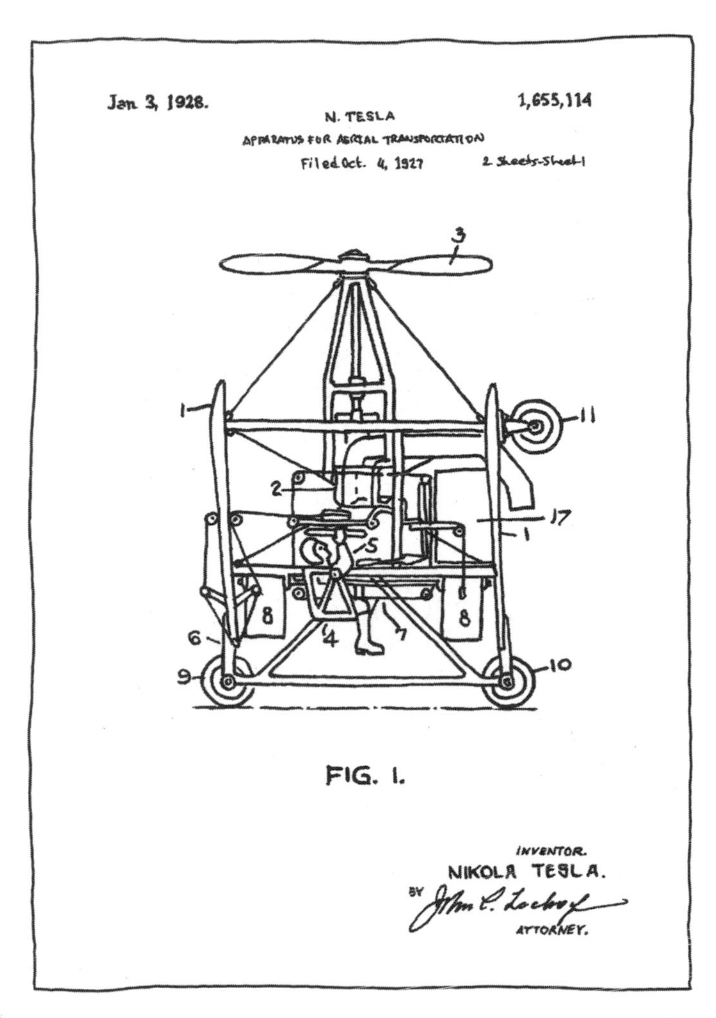

As with so many other of his ideas, Tesla was stymied in his efforts to raise money for a prototype of his "helicopter-plane." The patent was granted on January 3, 1928, but it would be his last. Though he never lived to see it, his aircraft was a precursor to today's VTOLs (vertical takeoff and landing planes). Eventually helicopters and a vast array of military fighter planes would be built as initially suggested by Tesla.

Hovercraft

Tesla's horseshoe-crab-shaped VTOL is said to have resembled a Corvette that rose up and road along the ground on a layer of air. The idea inherent in this invention found its way into today's Harrier jump jet fighter planes and air-cushioned vehicles (ACVs), or hovercraft, used to transport people and equipment over bodies of water, including the English Channel.

Radar

Tesla's vision did not stop with aircraft, of course. His mind was never at rest, inspired by changing needs as much as by spontaneous invention. The outbreak of World War I in Europe compelled him to act in the cause of dual patriotism. He was a U.S. citizen but still proud of his Serbian background and wished to contribute to the efforts against the aggressor Austria-Hungary. As a pacifist, he felt that any invention should be a means for ending war.

Relying on his previous discovery of stationary waves, Tesla conceived a means of locating German U-boats (submarines) with a technology that anticipated underwater radar. He made the following prediction in the August 1917 issue of *Electrical Experimenter*:

> *Now we are coming to the method of locating such hidden metal masses as submarines by an* electric ray.... *That is the thing which seems to hold great promises. If we can shoot out a concentrated ray comprising a stream of minute electric charges vibrating electrically at tremendous frequency, say millions of cycles per second, and then intercept this ray, after it has been reflected by a submarine hull for example, and cause this intercepted ray to illuminate a fluorescent screen (similar to the X-ray method) on the same or another ship, then our problem of locating the hidden submarine will have been solved.*

Insurmountable technical problems prevented the development of Tesla's system for the time being, but by World War II the military had developed sonar detection methods that were a viable alternative to Tesla's electric wave theory.

Bankruptcy

With all of his discoveries during this period, Tesla maintained the outward appearance of a rich man. This, like his bon vivant persona, was contrary to the fact of the situation. Tesla borrowed money from J.P. Morgan's son Jack, but it was not enough to keep his enterprises afloat and he repeatedly failed to meet his debts. Tesla's finances were on a precipitous roller coaster during the years 1912–1916.

His public image and need for funding demanded that Tesla move his office to a prestige location. He chose the Woolworth Building, then the tallest building in the world. He moved into the space in 1914 but was unable to pay the rent and moved to 8 West

40th Street, across the street from the New York Public Library and Bryant Park.

Tesla's finances came unraveled when he was summoned to court in 1916 for his inability to pay the city $935 in personal taxes. Tesla revealed that he lived at the Waldorf Astoria, but mostly on credit. His company had no assets, and he received only enough royalties on his patents to pay for basic expenses. He had scores of debts and more than five proceedings against him still pending. He was owed no money, had nothing in the bank, and had no personal assets. The court appointed a receiver to handle his financial affairs.

He was publicly disgraced, his credibility called into question.

If bankruptcy wasn't Tesla's final indignation, the dynamiting of his beloved Wardenclyffe Tower on July 4, 1917, precipitated another near breakdown. Rumors abounded that the U.S. Government razed the tower under the suspicion that it was being used to send messages to German U-boats off the coast. The real reason was that the mortgage holder, Waldorf Astoria proprietor George C. Boldt, needed to destroy the tower in order to make the property more saleable. The tower's remains were salvaged for scrap.

With many of his closest friends and associates dead by this time, Tesla was well on his way to becoming a recluse. His feelings toward humanity would only harden.

Nobelists

When Marconi first sent a wireless telegraph signal across the Atlantic Ocean in 1901, Otis Pond, an engineer then working for Tesla, commented, "Looks as if Marconi got the jump on you." Tesla replied, "Marconi is a good fellow. Let him continue. He is using seventeen of my patents." Marconi would go on to win the 1909 Nobel Prize in Physics (jointly with Karl Ferdinand Braun) for his contributions to the development of wireless telegraphy.

Tesla would never get over being usurped by Marconi. On August 4, 1915, Tesla sued the Marconi Corporation for copyright infringement but was financially unable to pursue the claim. Adding insult to injury, the *New York Times* on November 7 inexplicably reported that Tesla and Edison were to share the 1915 Nobel Prize. Tesla, who had not received official confirmation, was quoted as follows:

> *I believe that ultimately all battles, if they should come, will be waged by electrical waves instead of explosives.... I have concluded that the honor has been conferred upon me in acknowledgment of a discovery announced a short time ago which concerns the transmission of electrical energy without wires.*

Edison and Tesla To Get Nobel Prizes

Tesla, the article went on, "said he thought Mr. Edison was worthy of a dozen Nobel Prizes" but "knew new nothing of the discovery that induced the authorities in Sweden to confer the great honor on Mr. Edison." With his customary irony, Tesla was actually saying that Edison was not in his league for discovery. Robert and Katharine Johnson expressed congratulations, but there was little elation on Tesla's part. If there was, it was dashed a week later when the Royal Swedish Academy of Sciences announced that William Henry Bragg and his son, William Lawrence Bragg, had in fact won the year's Nobel Prize for Physics.

This was not the last time Tesla would be snubbed by the Nobel Committee, while at least four other scientists were awarded the prize in Tesla's lifetime for discoveries that were initially his.

YEAR	RECIPIENT	FIELD	ACHIEVEMENT
1927	Arthur Holly Compton	Physics	Proved the existence of cosmic rays consisting of high-velocity particles of matter
1935	Jean Frédéric Joliot	Chemistry	Discoverer of artificial radioactivity
1936	Victor Franz Hess	Physics	Discoverer of cosmic rays and a pioneer in radiation
1936	Ernest Orlando Lawrence	Physics	Developed the cyclotron (or as we know it—an "atom smasher"— Tesla's molecular-bombardment lamp, i.e. his carbon-button lamp)

The Edison Medal

While neither Tesla nor Edison would ever win a Nobel Prize, the American Institute of Electrical Engineers (AIEE), perhaps to make amends for the Swedish Academy's oversight, nominated Tesla for its highest award, the Edison Medal, in 1916.

B.A. Behrend, a fellow engineer and longtime champion of Tesla's AC theories, brought him the good news and was designated as the presenter of the award at a forthcoming AIEE banquet. Tesla at first refused Behrend's overtures, seething at the indignities that been heaped upon him for decades. The medal, he felt, was small consolation.

After countless entreaties, Tesla finally acquiesced. Attired in white tie and tails, he attended the banquet in his honor at the Engineers Club in Manhattan on May 19, 1917, addressing the attendees in his customary joking, charming manner. But when it came time to cross the street for the formal award presentation at the United Engineering Societies building, Tesla was nowhere to be found. After a frantic search, Behrend finally found him in nearby Bryant Park, along 40th Street, still in formal attire and surrounded by a crowd. His arms were raised like Saint Francis of Assisi, draped with pigeons. With a silk handkerchief drawn from his jacket pocket, he gently shooed away the pigeons and strode with Behrend to the auditorium to accept the Edison Medal.

Tesla gave a resounding, circumlocutory speech on Edison's virtues and described his own machine for changing the weather. Then, inexplicably, Tesla announced that his dream of wireless energy transmission had become a reality. "Recently," he said, "I have obtained a patent on a transmitter with which it is practicable to transfer unlimited amounts of energy to any distance."

It would not be the first time or last time that Tesla was lost among the pigeons. Strange smells were rumored to be emanating from his hotel room, with pigeon droppings lining the window sills. Tesla had also been spotted feeding pigeons at the nearby New York Public Library and St. Patrick's Cathedral.

Chapter 9
ENDINGS

KATHARINE JOHNSON'S DEATH IN 1925 contributed further to Tesla's estrangement from society. Her letters had given him a sense of connection and being cared for; now he retreated into his communion with pigeons. According to friend and biographer John O'Neill, he singled out one in particular:

> *a beautiful bird, pure white with light gray tips on its wings; that one was different. It was a female. I would know that pigeon anywhere.... No matter where I was that pigeon would find me; when I wanted her I had only to call her and she would come flying to me. She understood me and I understood her.... I loved that pigeon, I loved her as a man loves a woman, and she loved me... As long as I had her, there was a purpose in my life.* (*O'Neill,* Prodigal Genius: The Life of Nikola Tesla)

But even the pigeons would become a cause of consternation. Not only would Tesla have to move from hotel to hotel for not paying his bills, but the accumulation of pigeon excrement around his rooms motivated hotel proprietors to evict him. His possessions—dozens of

trunks containing documents, correspondence, theoretical papers, and invention models—were carted from one storage facility to another.

Tesla valued his privacy and could be curt with hotel employees, ordering them to remain at least three feet from him. Then, broke but contrite, he would generously tip hotel staff in order to make amends for his brusk manners.

Heralding the Future

With his days as an inventor behind him, Tesla found solace in his pigeons and indulged his eccentricities. He had always been a fastidious dresser, but by the 1930s his clothes were largely out of fashion, a throwback to another era. His fear of germs became the stuff of gossip. Meanwhile, Tesla was living off a trickle of royalties from the speedometer, some motors, and published articles. Eventually the Westinghouse Corporation agreed to pay him a monthly consulting fee and cover his hotel rent.

Tesla was back in the spotlight on July 20, 1931, when his picture graced the cover of *Time* magazine to mark his 75th birthday. The accompanying article set the tone for greeting reporters who would flock to him on subsequent birthdays. Tesla did not disappoint, making astonishing predictions for the future.

Pigeons were not the only flights of fancy for the aging Nikola Tesla. He continued to believe that his inventions could serve the good of all humankind. He believed that being able to communicate with pigeons was of a higher order. Similarly he felt that when he signaled other planets he reached other life forms. If he were able to signal other planets, he believed then he should be able to signal humans globally in the same fashion. Communication distances between all inhabitants would be shortened. With improved communication, the world would be a better place. Differences between peoples could be overcome. As a man of peace he believed wars could be avoided through improved communication

Transmitting wireless energy across the solar system held a grip on his thinking. He issued pronouncements on cosmic rays, transmitting mechanical energy, and particle beam weaponry, reasoning that if he could develop inventions in these areas then wars would no longer be necessary. Machines would settle any differences. From the 1930s to his death, Tesla claimed to have invented a "peace ray," a particle-beam projector he called a *teleforce.* The fanciful weapon was thought of instead as a "death ray," a reputation he attempted to dispel in an interview for an October 1934 article in *Every Week* magazine, "Dr. Tesla Visions the End of Aircraft in War."

The beam, intended chiefly for defense, will be projected from an electric power plant, ready to be put in action at the first sign of danger. The cost of operation will be insignificant, as the plant is

chiefly intended for use in times of peace ... nothing in common with the so-called "death ray."... It is impossible to develop such a ray. I worked on that idea for many years before my ignorance was dispelled and I became convinced that it could not be realized.

Did even a design for such a weapon exist? The U.S. military did not investigate further, but the FBI was fearful and went to great pains to confiscate Tesla's papers upon his death. After ten years of scrutiny, the papers were shipped to Belgrade under the auspices of Tesla's nephew, Sava Kosanović, the U.S. ambassador to Yugoslavia. The material is now housed at the Nikola Tesla Museum in Belgrade.

No "death ray" was ever found.

More Speculation

Since Tesla worked in a shroud of secrecy, he has left a legacy of mystery. Even if there was no death ray, other inventions are rumored to have resulted from his particle beam research. For example, Tesla professed to have created a secret apparatus that could produce energy in free air, thus eliminating the need for a high vacuum. This would also enable him to produce an electrical force of immense power, amplify that force, and ultimately produce a tremendous electrical repelling force.

On the occasion of his birthday throughout his remaining years, Tesla would make predictions and offer insights that bordered on—or clearly entered—the realm of the absurd.

- Human beings were really automatons under the control of outside forces. There was no individuality.
- Long life could be achieved on a meager diet and little or no sleep. Tesla's diet consisted primarily of bread and warm milk, with rigorous daily exercise as part of his regime.
- Someday it would be possible to photograph human thoughts. If a thought is reflected on the retina, then it would be possible to project the thought on a screen.

In the meantime, the world still awaited his apparatus that would transmit mechanical energy to any part of the globe and his atom smasher that would produce inexpensive radium.

Final Years

In 1936, to commemorate his eightieth birthday, the Yugoslavian government founded the Tesla Institute in Belgrade for the purpose of conducting applied research in electrical engineering.

The following year, Tesla would be honored by the Yugoslavian and Czechoslovakian governments with the Order of the White Eagle and the Order of the White Lion, respectively.

The celebrated inventor would also receive an annual monetary honorarium from the Yugoslavian government during his final years. In March 1937, meanwhile, he was elected to the Serbian Royal Academy of Sciences.

The year 1937 also marked a major turning point in Tesla's personal life. His friend Robert Underwood Johnson passed away that October after a series of illnesses. With both Katharine and Robert now gone, Tesla was all but alone with his pigeons. That August, departing the Hotel New Yorker late one night to feed pigeons in the park, Tesla was hit by a taxi and thrown across the street. Refusing medical treatment, he limped back to the hotel and engaged a

messenger to feed the pigeons until he could resume his routine. Tesla's health declined in the succeeding months, as he grew gaunt from his limited diet and undiagnosed ailments that may have resulted from the accident.

Tesla's withdrawal from society and failing health would become even more marked during the early 1940s. He grew increasingly paranoid about germs and became ever weaker and more emaciated. He took to bed sometime in 1942 and died quietly in his sleep, alone and penniless, on January 7, 1943. The next day, a maid ignored the "Do Not Disturb" sign on his door and discovered Tesla's body in Room 3327 of the Hotel New Yorker.

Afterwords

Before he could receive a funeral, the FBI—under the pretense that Tesla had been conducting espionage for Germany—ordered the seizure of all his belongings from the several hotels in which he had lived. The entire estate would remain in storage for almost a decade, until the FBI deemed it safe to be shipped to the Tesla Museum in Belgrade.

On January 12, 1943, more than 2000 people attended a state funeral, underwritten by the Yugoslav Government, at the Episcopal Cathedral of St. John the Divine in New York City. Mayor Fiorello La Guardia delivered a radio eulogy. Tesla's body was removed to a cemetery in nearby Ardsley, NY, where it would be cremated. Years later, his ashes were shipped to the Tesla Museum.

On June 21, 1943, in reversing an earlier patent ruling, the United States Supreme Court formally recognized Nikola Tesla as "the Father of Radio."

IT WOULD BE EASY TO END THE STORY HERE and say that Nikola Tesla was richly rewarded and highly honored—as well as maligned—both during his lifetime and after. Indeed he bore the mark of genius and left the world many gifts. He was a discoverer and an inventor who towered over the likes of Thomas Edison and Guglielmo Marconi, who have been more lionized. Even Edison's most famous invention, the incandescent light bulb, has been superseded by Tesla's fluorescent bulb and, more recently, programmable LED bulbs.

Tesla did not come to the United States and pursue his dreams of invention for the purpose of getting rich. He was totally disinterested in money except as a means of financing his research. His great early goal was to share his discovery of AC with the world and enable electricity to do the work of humanity. An innovator and iconoclast whose eccentricities we are now able to identify as obsessive compulsive disorders, Tesla exited the world with a record of achievement that is instrumental to all things electrical. He registered hundreds of patents but left few blueprints. A circle of mystery shrouds his life's work, but his legacy includes sustainable electric power, radar, radio, and wireless electrical power.

TESLA

Many of Tesla's predictions are the stuff of science fiction, contributing to his image as a cult figure, maverick, and untapped genius. Then again, many of his discoveries and inventions were so far ahead of their time that they seemed like science fiction at the time but later came to fruition. Tesla's patents continue to be mined as the basis of cutting-edge inventions. His bold innovation and out-of-the-box thinking have inspired the likes of Bill Gates, Steve Jobs, and Elon Musk to name a few. Perhaps you, too, will find inspiration in his life and work.

Tesla's undaunted spirit remains very much alive today as scientists, inventors, and entrepreneurs forge ahead in areas he pioneered:

- The efforts of a Balkan firm to make playground equipment that generates electricity from children's play;
- A California couple's explorations in magnetic-field architecture and hoverboard technology;
- Work at Johns Hopkins University and MIT to develop a robotic arm controlled by a person's mind;
- Research by a Boston group in transcranial direct-current stimulation (tcDCS) to alter mood;
- Plans by two Russian physicists to build a prototype of Tesla's Wardenclyffe Tower using modern materials and advanced electronics;
- NASA's discovery of methane gas on Mars, suggesting the possibility of life;
- Advanced development of flying cars in the United States, France, and elsewhere.

Stay tuned for the next invention inspired by Nikola Tesla!

FURTHER READING

Carlson, W. Bernard. *Tesla: Inventor of the Electrical* Age. Princeton, NJ: Princeton University Press, 2013.

Cheney, Margaret. *Tesla: Man Out of* Time. New York: Dell, 1983.

Cheney, Margaret, and Robert Uth. *Tesla: Master of Lightning.* New York: Barnes & Noble, 1999.

Hamilton, Tyler. *Mad Like Tesla.* Toronto: ECW, 2011.

Hunt, Inez and Wanetta W. Draper. *Lightning in His Hand: The Life Story of Nikola Tesla.* Hawthorne, California: Omni, 1981.

Jonnes, Jill. *Empires of Light: Edison, Tesla, Westinghouse, and the Race to Electrify the World.* New York: Random House, 2004.

Kent, David. *Tesla: The Wizard of Electricity.* New York: Fall River Press, 2013.

Martin, Thomas Commerford. *The Inventions, Researches, and Writings of Nikola Tesla.* New York: The Electrical Engineer, 1894; reprinted by Barnes & Noble, 2014.

McNichol, Tom. *AC/DC.* San Francisco: Jossey-Bass, 2006.

O'Neill, John J. *Prodigal Genius: The Life of Nikola Tesla.* New York: Ives Washburn, 1944; reprinted by David McKay, 1964.

Seifer, Marc J. *Wizard: The Life and Times of Nikola Tesla.* New York: Citadel, 1998.

Tesla, Nikola. *Lectures, Patents, Articles.* Belgrade: Nikola Tesla Museum, 1956. Mokelumne Hill, CA: reprinted by Health Research, 1973.

Tesla, Nikola. *My Inventions: The Autobiography of Nikola Tesla.* Edited by Ben Johnston. New York: "Electrical Experimenter," 1919: reprinted in Williston, VT: Hart Brothers, 1982.

Tesla, Nikola. *My Inventions and Other Writings.* New York: Penguin, 2011.

Tesla, Nikola. *The Problem of Increasing Human Energy.* Minneapolis, MN: Filiquarian, 2007.

Tesla, Nikola. *Very Truly Yours, Nikola* Tesla. Radford, VA: Wilder, 2007.

WEBSITES

Nikola Tesla Museum, Belgrade: www.tesla-museum.org/meni_en.htm

Tesla Memorial Society of New York: www.teslasociety.com

Tesla Science Center at Wardenclyffe: www.teslasciencecenter.org

Tesla Science Foundation: http://teslasciencefoundation.org

Tesla Universe: http://www.teslauniverse.com

ACKNOWLEDGMENTS

I AM INDEBTED TO NIKOLA TESLA for the wealth of information found in his own voluminous writings and patents, as well as www.teslauniverse.com.

Tesla's story has been told, retold, adapted, fictionalized, and factualized by many writers whose inspiration I am grateful for in my own research. Standouts are his biographers, Bernard W. Carlson, Margaret Cheney, Jill Jonnes, Thomas Commerford Martin, John J. O'Neill, and Marc J. Seifer. Without their scholarly enthusiasm it would be impossible to bring Tesla's life to a wider audience.

Any biographical journey begins somewhere, and for that I thank Godfrey P. Jordan who ignited the spark. Along the way I am grateful for those who offered much needed encouragement, suggestions, and advice: Bill Cole, Steve Dalachinksy, Nicholas Kosanovich, Bill Laswell, Bill Morgan, Lawrence D. "Butch" Morris, Sarah Sully, William H. Terbo, Charles Tyler, Vivek Tiwary, and Yoko Yamabe.

Additional gratitude goes to the team at *For Beginners,* especially Dawn Reshen-Doty, Publisher; Merrilee Warholak, Editorial Director; and Jeff Hacker, Acquisitions Editor, who masterfully streamlined the telling of Tesla's story, as well as Owen Brozman for his illuminating illustrations.

When it comes to keeping the electricity flowing, the greatest thanks goes to my loving wife Patricia, who, red pen in-hand, would check my meanderings, question my choices, and always be there with positive support.

ABOUT THE AUTHOR

Robert I. Sutherland-Cohen, a polymath and long-time Tesla enthusiast, is associate professor (emeritus) of stage management and production manager at Brooklyn College in New York. He has worked as a stage manager for some of America's most prestigious companies on Broadway, off Broadway, and in regional theater, as well as the New York City Opera. He has written extensively on stage production and is a widely published jazz performance photographer as well. Robert has a B.S. in mathematics from Northeastern University and an M.F.A. in directing from Boston University.

ABOUT THE ILLUSTRATOR

Owen Brozman is the illustrator of the *New York Times* best-seller *You Have to Fucking Eat* (Akashic, 2014); the acclaimed graphic novel *Nature of the Beast* (Soft Skull Press, 2009); and, in this series, *Ayn Rand For Beginners* (2009). Other publications and clients include *National Geographic*, *Scholastic*, TBWA\Chiat\Day, *Time Out New York*, Ninja Tune Records, and many others. Owen lives in Brooklyn, NY, with his wife and daughter. You can see more of his work at www.owenbrozman.com.

THE FOR BEGINNERS® SERIES

ABSTRACT EXPRESSIONISM FOR BEGINNERS:	ISBN 978-1-939994-62-2
AFRICAN HISTORY FOR BEGINNERS:	ISBN 978-1-934389-18-8
ANARCHISM FOR BEGINNERS:	ISBN 978-1-934389-32-4
ARABS & ISRAEL FOR BEGINNERS:	ISBN 978-1-934389-16-4
ART THEORY FOR BEGINNERS:	ISBN 978-1-934389-47-8
ASTRONOMYFOR BEGINNERS:	ISBN 978-1-934389-25-6
AYN RAND FOR BEGINNERS:	ISBN 978-1-934389-37-9
BARACK OBAMA FOR BEGINNERS, AN ESSENTIAL GUIDE:	ISBN 978-1-934389-44-7
BEN FRANKLIN FOR BEGINNERS:	ISBN 978-1-934389-48-5
BLACK HISTORY FOR BEGINNERS:	ISBN 978-1-934389-19-5
THE BLACK HOLOCAUST FOR BEGINNERS:	ISBN 978-1-934389-03-4
BLACK PANTHERS FOR BEGINNERS:	ISBN 978-1-939994-39-4
BLACK WOMEN FOR BEGINNERS:	ISBN 978-1-934389-20-1
BUDDHA FOR BEGINNERS	ISBN 978-1-939994-33-2
BUKOWSKI FOR BEGINNERS	ISBN 978-1-939994-37-0
CHICANO MOVEMENT FOR BEGINNERS:	ISBN 978-1-939994-64-6
CHOMSKY FOR BEGINNERS:	ISBN 978-1-934389-17-1
CIVIL RIGHTS FOR BEGINNERS:	ISBN 978-1-934389-89-8
CLIMATE CHANGE FOR BEGINNERS:	ISBN 978-1-939994-43-1
DADA & SURREALISM FOR BEGINNERS:	ISBN 978-1-934389-00-3
DANTE FOR BEGINNERS:	ISBN 978-1-934389-67-6
DECONSTRUCTION FOR BEGINNERS:	ISBN 978-1-934389-26-3
DEMOCRACY FOR BEGINNERS:	ISBN 978-1-934389-36-2
DERRIDA FOR BEGINNERS:	ISBN 978-1-934389-11-9
EASTERN PHILOSOPHY FOR BEGINNERS:	ISBN 978-1-934389-07-2
EXISTENTIALISM FOR BEGINNERS:	ISBN 978-1-934389-21-8
FANON FOR BEGINNERS:	ISBN 978-1-934389-87-4
FDR AND THE NEW DEAL FOR BEGINNERS:	ISBN 978-1-934389-50-8
FOUCAULT FOR BEGINNERS:	ISBN 978-1-934389-12-6
FRENCH REVOLUTIONS FOR BEGINNERS:	ISBN 978-1-934389-91-1
GENDER & SEXUALITY FOR BEGINNERS:	ISBN 978-1-934389-69-0
GREEK MYTHOLOGY FOR BEGINNERS:	ISBN 978-1-934389-83-6
HEIDEGGER FOR BEGINNERS:	ISBN 978-1-934389-13-3
THE HISTORY OF CLASSICAL MUSIC FOR BEGINNERS:	ISBN 978-1-939994-26-4
THE HISTORY OF OPERA FOR BEGINNERS:	ISBN 978-1-934389-79-9
ISLAM FOR BEGINNERS:	ISBN 978-1-934389-01-0
JANE AUSTEN FOR BEGINNERS:	ISBN 978-1-934389-61-4
JUNG FOR BEGINNERS:	ISBN 978-1-934389-76-8
KIERKEGAARD FOR BEGINNERS:	ISBN 978-1-934389-14-0
LACAN FOR BEGINNERS:	ISBN 978-1-934389-39-3
LIBERTARIANISM FOR BEGINNERS:	ISBN 978-1-939994-66-0
LINCOLN FOR BEGINNERS:	ISBN 978-1-934389-85-0
LINGUISTICS FOR BEGINNERS:	ISBN 978-1-934389-28-7
MALCOLM X FOR BEGINNERS:	ISBN 978-1-934389-04-1
MARX'S DAS KAPITALFOR BEGINNERS:	ISBN 978-1-934389-59-1
MCLUHAN FOR BEGINNERS:	ISBN 978-1-934389-75-1
MORMONISM FOR BEGINNERS:	ISBN 978-1-939994-52-3
MUSIC THEORY FOR BEGINNERS:	ISBN 978-1-939994-46-2
NIETZSCHE FOR BEGINNERS:	ISBN 978-1-934389-05-8
PAUL ROBESON FOR BEGINNERS:	ISBN 978-1-934389-81-2
PHILOSOPHY FOR BEGINNERS:	ISBN 978-1-934389-02-7
PLATO FOR BEGINNERS:	ISBN 978-1-934389-08-9
POETRY FOR BEGINNERS:	ISBN 978-1-934389-46-1
POSTMODERNISM FOR BEGINNERS:	ISBN 978-1-934389-09-6
PRISON INDUSTRIAL COMPLEX FOR BEGINNERS:	ISBN 978-1-939994-31-8
PROUST FOR BEGINNERS:	ISBN 978-1-939994-44-8
RELATIVITY & QUANTUM PHYSICS FOR BEGINNERS:	ISBN 978-1-934389-42-3
SARTRE FOR BEGINNERS:	ISBN 978-1-934389-15-7
SAUSSURE FOR BEGINNERS:	ISBN 978-1-939994-41-7
SHAKESPEARE FOR BEGINNERS:	ISBN 978-1-934389-29-4
STANISLAVSKI FOR BEGINNERS:	ISBN 978-1-939994-35-6
STRUCTURALISM & POSTSTRUCTURALISM FOR BEGINNERS:	ISBN 978-1-934389-10-2
TONI MORRISON FOR BEGINNERS:	ISBN 978-1-939994-54-7
WOMEN'S HISTORYFOR BEGINNERS:	ISBN 978-1-934389-60-7
UNIONS FOR BEGINNERS:	ISBN 978-1-934389-77-5
U.S. CONSTITUTIONFOR BEGINNERS:	ISBN 978-1-934389-62-1
ZEN FOR BEGINNERS:	ISBN 978-1-934389-06-5
ZINN FOR BEGINNERS:	ISBN 978-1-934389-40-9

www.forbeginnersbooks.com

COLIN KAEPERNICK

CHANGE THE GAME

This book is dedicated to all young people who trust their power, walk in their own truth, and change the game.

—Colin Kaepernick

To my grandma, who loved me just as I am.

—Eve L. Ewing

For my parents, who opened up my world through books.

—Orlando Caicedo

Photos courtesy of Know Your Rights Camp

Library of Congress Control Number: Available

ISBN 978-1-338-78965-2

10 9 8 7 6 5 4 3 2 1 23 24 25 26 27

Printed in the U.S.A. 40

First edition, March 2023
Edited by Michael Petranek with Christopher Petrella
Lettering by Hassan Otsmane-Elhaou
Color by Bryan Valenza
Book design by Jeff Shake
Special thanks to Tony Ng and Nessa

COLIN KAEPERNICK

CHANGE THE GAME

A GRAPHIC NOVEL

EVE L. EWING
ORLANDO CAICEDO

CHAPTER 1

IN THE MIRROR

CALIFORNIA
SOME YEARS BACK
I have never liked to lose.
But that day... that day, I was losing.
In sports, there's nothing unusual about that. You win some, you lose some. If you don't know how to take a loss...
...and figure out for yourself what it means and how it informs your life...
COULD BE A HOMER, FOLKS! AND COLIN KAEPERNICK IS UNDER SO MUCH PRESSURE...HE IS DEVASTATED.
You're not going to make it very far...
AND THERE IT GOES: A HOME RUN. THAT'S FIVE EARNED RUNS CHARGED TO KAEPERNICK THIS INNING.
ROUGH DAY FOR THIS TALENTED YOUNG PITCHER.
That day—that loss—would teach me a lesson I've never forgotten.

ALL RIGHT, COLIN, YOU'RE AT **RIGHT FIELD** THIS INNING. JASON, YOU'RE ON THE MOUND.
GOT IT, COACH.
YEAH... **GOT** IT.

YO, ARE YOU REALLY WALKING THROUGH OUR WARM-UP RIGHT NOW?
Colin— we're not supposed to do that.

TOUGH **GAME**, HUH, TERESA?
UH-HUH.

COLIN, WHAT'S UP?
MY BAD. MY SHOE JUST CAME **UNTIED**.

Later...
I KNOW YOUR PARENTS ARE SORRY THEY HAD TO MISS THE END OF THE GAME, SWEETIE.

YOUR DAD SAID HE HAD A WORK DINNER? HE'S REALLY MAKING MOVES OVER THERE AT THE PLANT.
Mm-hmm.

I KNOW YOU'RE UPSET, COLIN. BUT EVERYONE HAS A BAD DAY SOMETIMES.
Yeah. Thanks.

HELLO? ANYBODY HOME?
The note my father left me that day has stuck with me ever since.

"When you're trying to correct a problem, you should start by looking in the mirror."
I was so mad at him in that moment. Mad at him for putting me on the spot.

Mad at him because I knew he was right.

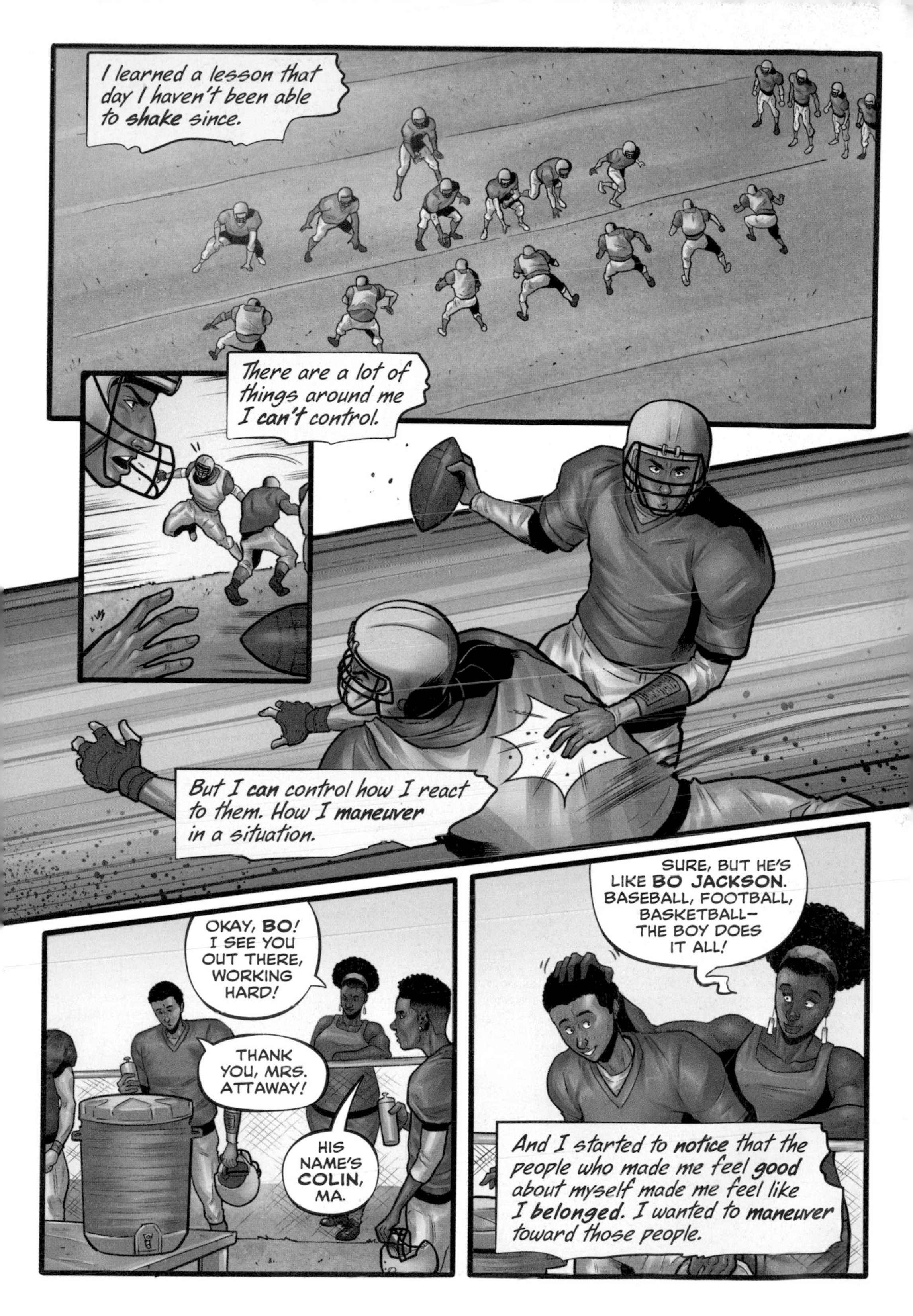
I learned a lesson that day I haven't been able to **shake** since.
There are a lot of things around me I **can't** control.
But I **can** control how I react to them. How I **maneuver** in a situation.
OKAY, **BO**! I SEE YOU OUT THERE, WORKING HARD!
THANK YOU, MRS. ATTAWAY!
HIS NAME'S **COLIN**, MA.
SURE, BUT HE'S LIKE **BO JACKSON**. BASEBALL, FOOTBALL, BASKETBALL— THE BOY DOES IT ALL!
And I started to **notice** that the people who made me feel **good** about myself made me feel like **I belonged**. I wanted to **maneuver** toward those people.

Look at him. He's hypnotized.
He's a teen boy, Rick. He's gonna have heroes.
Sure, but does it have to be...
ALLEN IVERSON!
ANSWER
IVERSON FOR THREE!
A.I. WITH THE CROSS-OVER! WOW!
I worshipped Iverson. Not just because he was an amazing ballplayer. It was everything about him. His drive. His style.
HERE'S IVERSON IN THE OPEN COURT!
A.I. BEHIND THE BACK! AND THAT'S WHY THEY CALL HIM THE ANSWER!
He wore his Blackness like a suit of armor.
And I wanted to be that way. To feel that way.

SEE YOU LATER, MOM.

WHAT'S UP, RYAN?
HEY, BRO.

HEY, SO...I was wondering... WHERE DID YOU GET YOUR **HAIR** DONE LIKE THAT?
WHAT, YOU MEAN CORNROWS?
YEAH. **CORNROWS.**

LET'S SEE WHAT YOU **GOT** HERE.
YEAH, THIS'LL WORK. MY BROTHER HAS A FRIEND. SHE CAN HOOK YOU UP.
BUT YOU GOTTA GROW IT **OUT** FIRST. ABOUT **THREE INCHES** SHOULD GET YOU RIGHT.

THREE INCHES. GOT IT. **THANKS.**
NO **PROBLEM,** MAN.

READY TO HEAD OUT, BO?
I NEED TO TAKE THE CAR FOR AN OIL CHANGE, BUT WE SHOULD BE ABLE TO SQUEEZE YOUR **HAIRCUT** IN FIRST.
MM, NO THANKS. I DON'T NEED A **HAIRCUT** TODAY.

SECOND SATURDAY IS **SUPERCUTS SATURDAY**, COLIN.

I'VE DECIDED I'M **GROWING** MY **HAIR** OUT.
YOU'RE WHAT?!

AND I'M GETTING **CORNROWS**.

He's getting **what** rolls?

I didn't have the words then for what I was experiencing.

I just saw the stares. Heard the whispers.

ORGANIC
He looks like a CLOWN.
And I got the message loud and clear...

My existence, my body, the way my own hair grew out of my head naturally...
That was somehow not okay.

Eventually, they got tired of arguing with me.
WELL... THIS IS IT. JUST CALL ME WHEN YOU'RE DONE, I GUESS.

YOUR HAIR IS REALLY DRY.
WAS I SUPPOSED TO WET IT?
What?
Oh...you really don't... NO, WHAT I MEAN IS, YOU HAVE TO MOISTURIZE YOUR HAIR.
I'LL GIVE YOU SOME PINK OIL TO TAKE WITH YOU.
AND I DON'T KNOW IF THIS IS LONG ENOUGH. YOU DON'T WANT TO COME BACK IN A MONTH?
A MONTH?!
IT'S FINE, I GUESS. I CAN USE RUBBER BANDS. I'LL FIGURE SOMETHING OUT.
This was it, I knew it. The beginning of a new journey. I would be a new person. I could see it all now, the path laid before me...

DEEDEE, I HAVE AN EMERGENCY!
THE MALL IS ABOUT TO CLOSE, AND I NEED THAT TOP THAT WAS ON SALE.
TASHA, THAT'S NOT AN EMERGENCY.
IT IS! WHAT AM I SUPPOSED TO WEAR TO THE BASKET-BALL GAME?

THIS'LL JUST TAKE A SEC. THEN WE'LL GET BACK TO THE HOUSE AND I'LL FINISH YOU UP.
UH-UH! WAIT IN THE CAR. LOOK AT YOUR HEAD! I HAVE A REPUTATION TO PROTECT OUT HERE.
I'LL LEAVE THE AC ON.

THAT WAS QUICK.
TOLD YOU! WE NEED TO MAKE ONE MORE STOP, THOUGH.
WE RAN INTO MARCUS, AND HE SAW THAT THE KRISPY KREME HOT LIGHT IS ON.
WHAT'S KRISPY KREME?

IT'S A NATURAL THING, AND I'M SMOOTH AS CREAM...
AND I KNOW YOU CAN SELL A FOOL A DREAM!
At this point, this was shaping up to be the greatest day of my life.
Uh-oh.
MOM

WE WANT TO SUPPORT YOUR DECISIONS, AND OF COURSE IT'S YOUR HAIR. BUT WE JUST DON'T WANT YOU TO LOOK LIKE...
...Unprofessional.
...LIKE A LITTLE THUG.
YOU NEED TO CUT YOUR HAIR.

I didn't care what my parents thought. I knew what I thought of myself, and the corners of the world where I could be my full self.
And I knew how I had felt in the tiny world of that kitchen, when DeeDee was doing my hair. I felt like my full self.
And it felt really, really good.
There was only one minor technical problem.

SO IF THIS SIDE OF THE TRIANGLE IS X, AND THIS SIDE IS THREE INCHES, HOW WOULD WE WRITE A FORMULA TO DETERMINE...

DeeDee's attempts to make my barely-long-enough hair work with the rubber bands was having some **consequences** I had **not** planned for.

I MEAN, YOU TOLD ME IT WOULD **HURT**, BUT THIS IS LIKE... IT HURTS **A LOT**.
OKAY, LET ME TAKE A LOOK.

AHA. SHE HAD TO USE THESE TIGHT LITTLE RUBBER BANDS? THIS LOOKS ROUGH, BRO. I THINK YOU NEED TO TAKE THESE **OUT**.
TAKE THEM **OUT**?!
HAS TO BE **DONE**. YOU LOOK LIKE YOU GOT A **FACE-LIFT**. I **KNOW** YOUR **FOREHEAD** HURTS RIGHT NOW.

AFTER I DRIVE TO THE BACK SIDE OF **WHO KNOWS WHERE** TO GET HIS HAIR DONE, THEN HE'S OFF TO THE **MALL**, WITHOUT PERMISSION, THEN I HAVE TO **WAIT** AN EXTRA **HOUR** FOR HER TO **FINISH**...
NOW HE WEARS IT FOR ONLY **ONE** DAY?
I DON'T CARE WHAT HE THINKS, RICK. HE'S CUTTING HIS HAIR.
I didn't grow my hair out again after that. Not for a **long** time.

CHAPTER 2

A REAL PROFESSIONAL

SENIOR YEAR
KAEPERNICK ON THE MOUND...

Eventually I learned...

To focus.
23
23
PITMAN
To tune out all distractions.

AND WITH THAT, KENNEDY IS OUT! WHAT A GAME SO FAR FOR PITMAN AND FOR COLIN KAEPERNICK.
SWING AND A MISS, STRIKE THREE!
SWING AND A MISS!
THAT'S ANOTHER STRIKEOUT!
SWING AND A MISS, AND KAEPERNICK HAS TWELVE STRIKEOUTS ON THE DAY!
NOW KAEPERNICK AT BAT...
KRAK
LOOK AT THAT!

COLIN—DAN DAVIDS, MODESTO BEE. WOW, WHAT A GAME!
YOU JUST PITCHED A SHUTOUT AND KNOCKED IN A RUN IN THE FOURTH INNING. HOW DO YOU FEEL?
PITMAN 23

I FEEL GOOD, BUT IT WAS A TEAM EFFORT. WE'VE BEEN PUTTING IN WORK, AND TODAY WAS THE REWARD.

GREAT GAME, COLIN!
JUST LAST WEEK, MERCED DELIVERED YOUR WORST LOSS OF THE YEAR. WOULD YOU SAY THIS IS A BIT OF A COMEBACK? OR EVEN VENGEANCE?

I JUST TRY TO TAKE EVERYTHING ONE DAY AT A TIME, ONE GAME AT A TIME, ONE INNING AT A TIME. THAT'S ALL I CAN DO.

ALL RIGHT, LAST QUESTION.
IT'S YOUR SENIOR YEAR. YOU'VE GOT SOME STAR POWER. WE'VE HEARD THAT YOU HAVE NOT ONLY DIVISION I INTEREST BUT ALSO MLB HAS BEEN POKING AROUND. CARE TO GIVE US ANY INSIDE SCOOP ON WHERE YOU'RE HEADED?

WHERE I'M HEADED...?
TO BE HONEST, MR. DAVIDS... I COULDN'T TELL YOU MYSELF AT THE MOMENT.

Thanks again—looks like you have other people waiting to talk to you.
P

PITMAN
23

Throughout my senior year, it felt rare to get an hour to myself. I was exhausted a lot, running ragged between games, working out, college recruitment, and keeping up my grades.

HOW YOU DOING, BO? WHAT A GAME THAT WAS. I CAN'T BELIEVE BEN PULLED OFF THAT DOUBLE PLAY.
YEAH, IT WAS A GOOD GAME.

WHATCHA WATCHIN'?

MAN, WE PAY FOR CABLE JUST FOR THERE TO BE SO MANY COMMERCIALS.
WOW, LOOK AT HER. DO YOU THINK SHE'S PRETTY?

I DUNNO. NOT... not really.

Home should have been my safe haven...

...but that wasn't always the case.

JUST SOME MUSIC VIDEOS.

My dad used to ask me these weird questions about who on TV I found attractive... I guess you could say it was sort of a game.

HOW 'BOUT HER?
No...?

HEY, I THINK THE BREWERS GAME IS ON.
GRIFFEY UP NEXT. HE'S HAVING QUITE THE SEASON, CREEPING UP TOWARD MICKEY MANTLE FOR 12TH PLACE ALL-TIME IN CAREER HOMERS.
Pfft. MICKEY MANTLE HE IS NOT.
YOU KNOW, I WISH THESE GUYS REALIZED THAT BEING A PRO BALLPLAYER MEANS JUST THAT. YOU HAVE TO BE A REAL PROFESSIONAL.
YOU GET PAID LIKE A PROFESSIONAL, YOU ACT LIKE A PROFESSIONAL. ON AND OFF THE FIELD.
LIKE I'VE ALWAYS TRIED TO TEACH YOU.

My parents were making a lot of things **crystal clear** without ever really saying them, and it weirded me out.
I tried to **shut** it all **out** and focus on other things.
Everybody was concerned that I was **too skinny**. They worried that I would get **injured**.
I always told them: **"No one is going to hurt me."**
And I had plenty of **other** things to think about, too.
TAKE IT TO THE BARN, PUT IT UP FOR THE NIGHT, CLIMB UP IN THE LOFT, SIT AND TALK WITH THE RADIO ON.

My schedule was **stacked**. Advanced Math, Advanced English, History, Photography, then study hall, where I tried to get all my **homework** done before **practice**.

YADADAMEAN? FO' SHO. KNOW WHAT I MEAN? FO' SHEEZY. I DON'T PUT THAT ON, THAT'S MY WORD. HYPHY.

I was comfortable at school, and confident.
But I never felt like I was the one who had to be in the middle of the action.

I like being back against the wall, watching.
There was always so much happening. Everybody had their own little group.

And I had mine.
YO.
HEY, MAN.

Zack. Football, and football only. The guy lived and breathed for the game, especially on the strategy side. Dreamed of being a coach someday.
Adrian. Funniest person I ever met in my life. Played basketball and baseball but not football.
Devin. Football, basketball, and ran track. Good enough to play at a Division I school, but everybody expected he would go to Stanford and be an engineer instead.

Jeremy. Football, basketball. No baseball. In the spring, he was always in the school musical.
Then we had Marcus. He was my lineman in football and my catcher in baseball.
There was Miguel. He was year-round, like me. Any sport you ever heard of, he played it. Football, basketball, baseball.
And Shaun. Biggest, strongest dude in the school... and refused to play sports! He was really into music. He played five instruments. But he hung out with us anyway.

Most of these guys, I had known since I was eight or nine. Some of us played on travel teams together. We knew each other's families. Our stories.
Knew everything about each other, really.

AND LIL JON IS LIKE,
"YEEEYUH!"
BRO, STOP!
KOFF
koff
YOU MADE SHAUN LAUGH TOO HARD! LOOK AT HIM, HE'S CHOKING!

YOU COULD BE ON TV, I SWEAR.
I LIVE TO SERVE.
We didn't talk about race too much. Nothing serious, really. But it wasn't a coincidence that our group looked the way it did.
Without saying it out loud, we had formed a little bubble of protection.

WE DEFINITELY HAVE A QUIZ.

YOU SURE? WHERE'D YOU HEAR THAT?

I'M SURE. I HEARD IT FROM ELIJAH PHILLIPS. HE HAS HER FIRST PERIOD.
SHOOT. I WAS GONNA USE STUDY HALL TO WORK ON THIS STATS PROBLEM SET I HAVE DUE TOMORROW, BUT I GUESS I'LL BE BRUSHING UP ON TONI MORRISON.

THANKS, JEREMY.

DUDE, WHO WAS THAT?
Huh? OH, TIFFANY. TIFFANY SHIELDS. SHE'S IN MY ENGLISH CLASS.

Tiffany Shields.

HI!

SPLASH

SORRY TO STARTLE YOU. YOU'RE JEREMY'S HOMIE. JUST WANTED TO SAY WHAT'S UP. I'VE SEEN YOU DURING **BASKETBALL** PRACTICE.

YEAH.

WELL...SEE YOU **AROUND**, I GUESS.
NICE TO MEET YOU!

CHAPTER 3
WHERE WE BELONG

For a lot of people, senior year in high school is a chance to let loose.
For me, it was a chance to double down and focus on the things that I thought would matter for my future.
Modesto Be
No Days Off for Kaepernick
looking good! ♡ -Mom

And in football...
I WANT TO SEE TRUST. I WANT TO SEE LEADERSHIP. I WANT TO SEE FOCUS.
LET'S GO— PRIDE ON THREE! ONE... TWO... THREE—

This was supposed to be our year.
PRIDE!

The year the Pitman Pride could beat Tracy West and bring home the section title.

NINETY-NINE...!
This year meant everything to me.
I wanted to get out of Turlock more than anything, and I figured this was my way to do it. Leading my team to the title could open up a lot of opportunities for me.

THAT'S A HUNDRED. SHEESH, YOU'RE NOT PLAYING AROUND. HERE, MAN, TAKE A BREATHER–
I just knew that whatever I wanted, I would have to work for it.

YOU'RE UP!
ALL RIGHT, SPOT ME. WHEN YOU'RE DONE DRINKING.
MAN, KAP, LEAVE SOME FOR THE FISH!

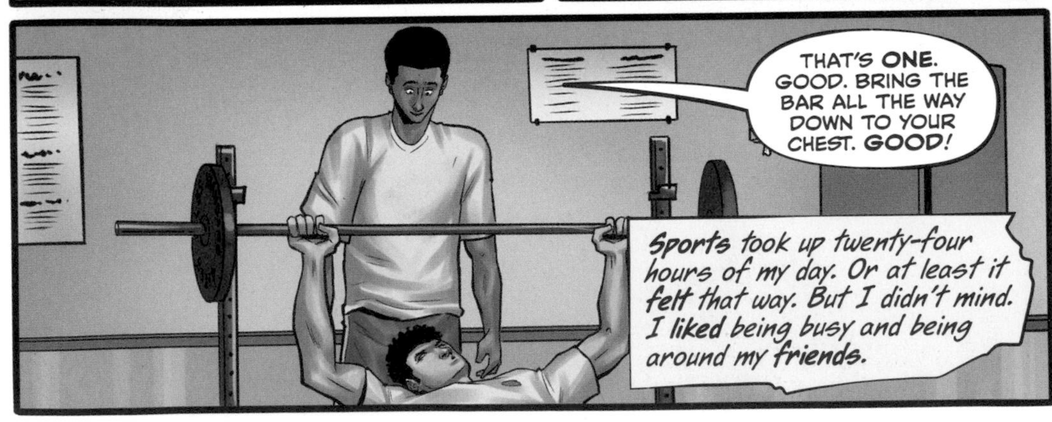
THAT'S ONE. GOOD. BRING THE BAR ALL THE WAY DOWN TO YOUR CHEST. GOOD!
Sports took up twenty-four hours of my day. Or at least it felt that way. But I didn't mind. I liked being busy and being around my friends.

!
I DIDN'T SEE IT YET. I MIGHT TRY TO GO FRIDAY.
Oh, MY BROTHER IS A HUGE BLADE FAN. HE'S GONNA DRAG US OUT THERE OPENING NIGHT.
Bro!

HOW MANY WAS THAT?
Ummmm...
WHAT HAPPENED TO SAFETY FIRST?

KAP, DON'T FEEL TOO BAD IF HE DROPS THAT THING ON HIS HEAD.
HE BARELY USES IT ANYWAY.

COLIN? EARTH TO KAEPERNICK! COME IN, KAEPERNICK!

I HONESTLY HAVEN'T EVEN BEEN TO THE MOVIES IN, LIKE, A MONTH. I'VE BEEN PULLING EXTRA HOURS AT THE STORE.
I DON'T KNOW HOW YOU HAVE TIME FOR A JOB AND SCHOOL AND PRACTICE.

Um...
SORRY TO INTERRUPT YOU. JUST WANTED TO SAY HI. I'M JEREMY'S FRIEND—

YEAH, COLIN!
I KNOW. BUT NICE TO FORMALLY MEET YOU.

I'D SHAKE YOUR HAND, BUT I'M A LITTLE SWEATY.
THIS IS MAKAYLA.
Hi.

WELL, BYE!
SEE YOU...

HOW DID THAT GO?
Huh? Oh.

FINE.

BACK TO WORK, CASANOVA.
WE NEED TO GET AT LEAST THREE MORE SETS IN BEFORE THE BELL.

Every time I saw Tiffany, she was smiling. She seemed to make everyone feel good about themselves.
SO THEN WHAT?
THEN WE CONVERT THIS TO SINE... AND THIS TO COSINE...
GOOD!
Including me.

THAT WAS A SMOOTH LITTLE CROSSOVER. I SEE YOU! A YOUNG IVERSON.
HA! NAH. I MEAN THANKS.
♪
Oops. THAT'S MY CUE.
But more than that, she had confidence. She seemed so comfortable in her own skin.
Laid-back. Chill.
It was like anywhere Tiffany went, she knew she was meant to be there.
I admired that.

I, on the other hand...

...distinctly did not feel that way.

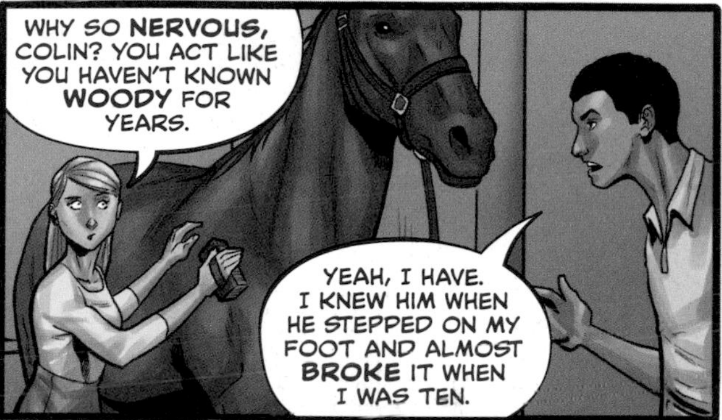
WHY SO NERVOUS, COLIN? YOU ACT LIKE YOU HAVEN'T KNOWN WOODY FOR YEARS.
YEAH, I HAVE. I KNEW HIM WHEN HE STEPPED ON MY FOOT AND ALMOST BROKE IT WHEN I WAS TEN.

PLUS, YOU CAN KNOW SOMEONE YOUR WHOLE LIFE AND NOT REALLY KNOW THEM, RIGHT?
Huh?

GOOD POINT, BOY. NOTHING WRONG WITH HAVING SOME HEALTHY FEAR. IT'S AN ANIMAL, AFTER ALL.

Can we go now, please?

GO, DEVON!

COME ON, COLIN, DON'T YOU WANT TO CHEER ON YOUR **SISTER?** SHE GETS NICE AND LOUD AT **YOUR** GAMES.

WOOOOOOOO!
GO, DEVON!

TWO **HOT DOGS,** PLEASE.
IS THAT **RICK KAEPERNICK** I SEE?

MIKE O'BRIEN! HOW **ARE** YA?!
I wanted the ground to open up...

...and swallow me whole.

In that moment, I felt like an alien from another planet.
Didn't my dad see what I saw? He had to.

COLIN?
COLIN?

SORRY, WHAT WAS THAT?
I SAID, ARE YOU GONNA BE GOING INTO THE FAMILY BUSINESS? A BIG MAN AT THE DAIRY PLANT LIKE YOUR DAD AND YOUR BIG BRO?
Even though I was in high school at the time, I still knew that the Confederate flag symbolized slavery and racism.

Nah, that's... not really for me.
DAD, I'M GONNA GO... TO... I'll be right back. I just... FORGOT SOMETHING.

COLIN, WAIT! WHERE ARE YOU— WHAT ABOUT YOUR HOT DOG?

A FEW DAYS LATER.
My dad was the one who drove me to practices and camps, so we spent hours and hours in the car together.
'CAUSE A MISSISSIPPI GIRL DON'T CHANGE HER WAYS JUST 'CAUSE EVERYBODY KNOWS HER NAME...

I always knew when a Big Talk was coming.

SON, I WANTED TO TALK TO YOU ABOUT THE OTHER DAY AT YOUR SISTER'S RODEO.
I JUST WANT YOU TO KNOW THAT... WELL...SOME PEOPLE IN TURLOCK...
How can I put this...?

YOU SEE, COLIN, THE MOST IMPORTANT THING THAT I WANT YOU TO KNOW IS THAT YOU BELONG. YOU–
WE'RE A FAMILY, ME, YOU, YOUR MOM, YOUR BROTHER AND SISTER. AND YOU BELONG. WE ALL BELONG.
AND–

COLIN?
Ah, poor kid is fast asleep. All he does is go from place to place, and he doesn't get enough rest.

♫ THE GEORGIA RAIN ON THE JASPER COUNTY CLAY COULDN'T WASH AWAY WHAT I FELT FOR YOU THAT DAY... ♪

♪ JUST YOU AND ME DOWN AN OLD DIRT ROAD... ♫

CHAPTER 4
FOCUS

BRETT FAVRE WITH TWO TOUCHDOWN PASSES HERE THIS AFTERNOON...
COLIN!
COLINNNN!
WE'RE ALL WAITING FOR YOU, BUDDY!
THROW IT HERE, COLIN!
LET'S SEE THAT FOCUS!

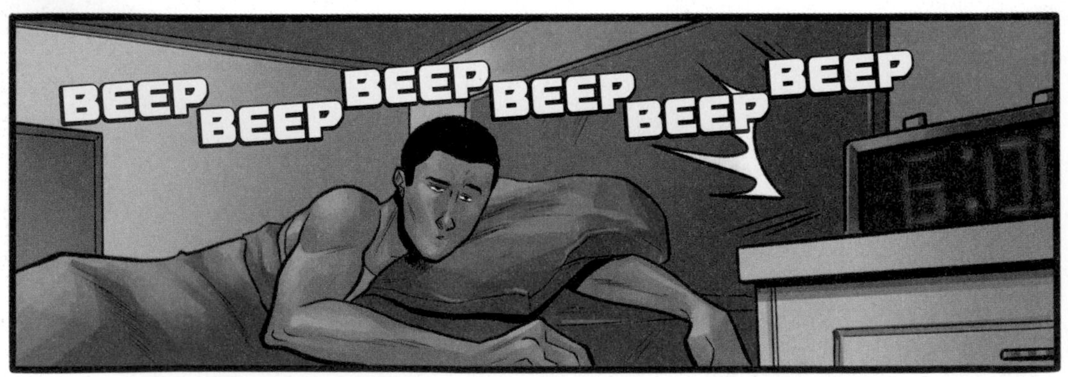
BEEP BEEP BEEP BEEP BEEP BEEP

MORNING, BO. HOW'D YOU SLEEP?

I SLEPT OKAY.
I HAD A REALLY **VIVID** DREAM.
ABOUT GRANDPA.

ABOUT GRANDPA... **WOW.** OKAY. ARE YOU HEADING OUT TO LIFT?
YEAH. I'M GONNA GO **WORK** OUT.
Oh.

Okay.

My grandpa was really **special** to me.

Not one second that I was with **him** did I ever question **who** I was or **what** I was.
I was his **grandson**, and that was **that**.

I took every opportunity possible to be places where I could let my guard down.
BURGERS
I had less and less patience for anything else.
BURGERS
PROBABLY BASKETBALL. I MEAN, THAT'S WHAT I'M WORKING TOWARD RIGHT NOW, YOU KNOW?
I KNOW THE ODDS ARE TOUGH. I MIGHT NOT MAKE IT AS A PRO. BUT I'VE BEEN AT IT SO LONG... IT'S HONESTLY HARD TO IMAGINE ANYTHING ELSE.
WHATEVER I DO... I REALLY JUST WANT TO BE THE BEST VERSION OF MYSELF.
DO YOU GET WHAT I MEAN?
I GET WHAT YOU MEAN.

AT FOUR?
SOUNDS GREAT.

BRIIIIIIING
SHOOT— I FORGOT MY HISTORY BOOK.
WELL, I GOTTA RUN. MR. S. DON'T PLAY WITH TARDIES. SEE YOU!

OKAY! OKAY! HERE'S ONE. HOW DO YOU STARVE A NIGGER?
HIDE HIS FOOD STAMPS UNDER HIS WORK BOOTS.
HA HA HA!

SLAM

The space my friends and I had built for each other... I knew then that it couldn't be the whole world. It was too good to be true.
My parents always told me...
NEVER EVER FIGHT SOMEONE UNLESS THEY ATTACK YOU FIRST AND YOU'RE FIGHTING BACK.
GH SCHO
What if things weren't so simple?

It seemed that everywhere I looked I was being ***bombarded*** *with the same old story.*

And ***that...*** *was beginning to feel like an* ***attack.***

IT'S A GLORIOUS DAY HERE AT DEDEAUX FIELD, WHERE THE BEAUTIFUL UNIVERSITY OF SOUTHERN CALIFORNIA IS WELCOMING SOME OF THE STATE'S **TOP** HIGH SCHOOL BASEBALL TALENTS FOR THE PUEBLO INVITATIONAL.

IT'S RARE TO SEE THIS MUCH TALENT IN ONE PLACE, FOLKS! WHAT AN OPPORTUNITY FOR THESE TOP-NOTCH PLAYERS.

Sometimes, one path seems easy. The sun is shining on it. It's neatly paved. You could just take that path, and go.
PITMAN
23

HE'S FIDGETING.
NO, HE'S NOT. HE'S READY FOR THIS.

But just because a path is easy, does that make it the right way to go?
What if there's something else waiting for you out there?

A path that's harder. A path that's maybe even concealed from you. But the right path.

On the mound in front of all these scouts, it felt like a Wild West showdown at times...

Scouts were holding their radar guns like they were in a Wild West shootout.

How would you know which road to take?

BAM
WHOOSH
SHWIFF
BAM
CLAPCLAP CLAPCLAP
PITMA
23
94...?!

!
!!!
PITMAN
23
HRRGH
EWIIP
SWIIIIF

I came away from that tournament with a **clear** feeling about **one** thing.
You might be **bigger.** But you're not more **skilled** than I am.
MAJOR TALENT, SON.
LET'S TALK—
A STANDOUT **SHOWING.**
THANK YOU. THANKS VERY **MUCH.**

IF YOU'LL **EXCUSE** ME... I JUST NEED TO GRAB SOME **WATER.**

Phew

Choices are never as easy as people make them out to be.
PITMAN
23

Sometimes there are signs, though.
THAT WAS QUITE THE PERFORMANCE, SON.
Oh. THANKS.
THE NAME'S ANDREW HAMILTON. UNIVERSITY OF TEXAS.

TEXAS!
THAT'S RIGHT. RANKED NUMBER ONE THIS YEAR.
I HEARD THERE'S A REALLY GOOD ECONOMICS MAJOR AT UT. AND HISTORY.
PITMAN
23

Huh? Oh, SURE. LISTEN, WE NEED SOMEONE WITH YOUR SHEER PHYSICALITY ON OUR TEAM.
I'M READY FOR A TOP-RANKED SCHOOL LIKE THAT. I ACTUALLY HAVE A 4.5 GPA, AND I'M INTERESTED IN–
YOU GETTIN' A HOT DOG, MITCH? I MIGHT JOIN YOU.

HELLO THERE! I'M RICK KAEPERNICK.
PITMAN
GLAD YOU'RE EXCITED TO MEET MY BOY!
MY SON HAS QUITE AN ARM, DOESN'T HE?
YOU'RE— Oh, HE—HE SURE DOES!
AND SMART AS A WHIP, TOO, IT SOUNDS LIKE!
NOW, I ALWAYS SAY, SHEER PHYSICALITY ISN'T ENOUGH TO REALLY SHINE. YOU NEED REAL INTELLIGENCE. LEADERSHIP ABILITY.
ABSOLUTELY. ISN'T THAT RIGHT, COLIN?
PITMAN 23

BIG STAR LIKE YOU, YOUNG MAN, HOW COME NOBODY LOCKED YOU DOWN DURING FALL SIGNING DAY?

WELL...
PIT

I'M SERIOUS ABOUT MY **PROSPECTS**. JUST NEED TO **THINK** SOME THINGS **THROUGH**.
PITMAN
23

WELL, DON'T THINK **TOO** HARD, SON! SPRING SIGNING DAY IS JUST AROUND THE CORNER.
I'LL BE IN TOUCH!

A few days later...
"JUST AROUND THE CORNER." LIKE I'M NOT THINKING ABOUT IT EVERY SECOND.
MAN, THAT'S A LOT OF PRESSURE.
SO WHAT'S STOPPING YOU FROM COMMITTING?
DON'T WORRY, I DON'T THINK ANY SPIES FOLLOWED US OUT HERE. YOUR SECRET IS SAFE WITH ME.

I don't really want to play baseball. I feel like... it's not for me.
BUT YOU'RE AMAZING AT IT.

THANKS.

IT'S JUST, SOMETIMES I FEEL LIKE I DON'T BELONG OUT THERE.

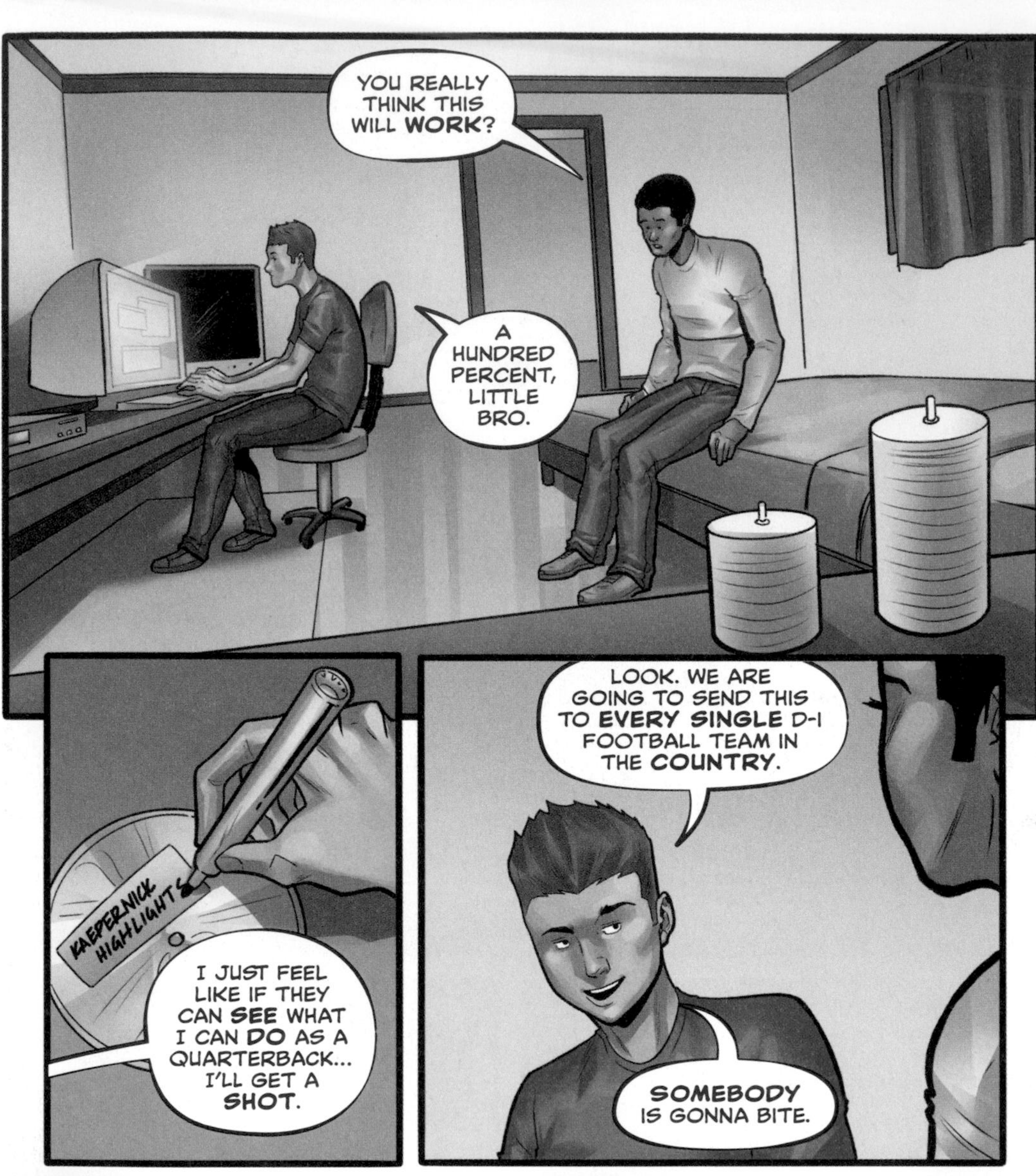
YOU REALLY THINK THIS WILL **WORK**?
A HUNDRED PERCENT, LITTLE BRO.
KAEPERNICK HIGHLIGHTS
I JUST FEEL LIKE IF THEY CAN **SEE** WHAT I CAN **DO** AS A QUARTERBACK... I'LL GET A **SHOT**.
LOOK. WE ARE GOING TO SEND THIS TO **EVERY SINGLE** D-I FOOTBALL TEAM IN THE **COUNTRY**.
SOMEBODY IS GONNA BITE.

WESTERN CAROLINA, WESTERN ILLINOIS, WESTERN KENTUCKY, WESTERN MICHIGAN...

DONE!

NO UNREAD

No bites?
TIFFANY

Nah.
Things will turn around with section playoffs. Watch.

ZZZZZ

CHAPTER 6
PLAYOFFS

It was almost time for section finals.
GO PITMAN! BEAT LINCOLN
And everyone was losing their minds.
PITMAN PRIDE!

THIS STUFF ALWAYS MAKES ME FEEL A LITTLE **WEIRD**.
I DUNNO. IT'S KINDA FUN TO BE A **CELEBRITY** FOR THE DAY.

PAM, WHERE'S THE CASHBOX FOR THE RAFFLE TICKETS?
HERE IT IS.

SO, WHAT DO YOU THINK OF OUR CHANCES THIS YEAR?
PITMAN PARENT BOOSTERS
OH, I THINK IT'S LOOKING **GOOD**. LOOKING REALLY GOOD.

THAT **KAEPERNICK**... HE'S SO **PHYSICAL** FOR A QUARTERBACK. YOU KNOW?
LIKE, **GREAT** STRENGTH FOR A KID THAT AGE.
GO PRIDE

IF YOU GET TO TALK TO **COLIN**, HE MIGHT **SURPRISE** YOU. REALLY ARTICULATE KID. SMART FOR AN **ATHLETE**.
BUT YEAH, WE HAVE A FEW **MONSTERS** THIS YEAR.

"I MEAN, THAT COLIN KID, HE'S JUST AN ANIMAL OUT THERE.
"AND THEN YOU GOT WHAT'S HIS NAME, NUMBER SIX? DEVIN? HE'S A SILVERBACK GORILLA.
"DON'T FORGET ZACK. HE'S A REAL BEAST, TOO. THAT KID IS SCARY!
31
"MARCUS—NOW THERE'S A LEADER. TALK ABOUT HIGH FOOTBALL I.Q."

I just knew it felt wrong. Later, I came to understand that describing Black athletes as animals or beasts portrays them as subhuman and less intelligent than white players.

But I knew that beating Lincoln would take more than words.

Our key word was discipline. We knew the Lincoln guys were way bigger than us.

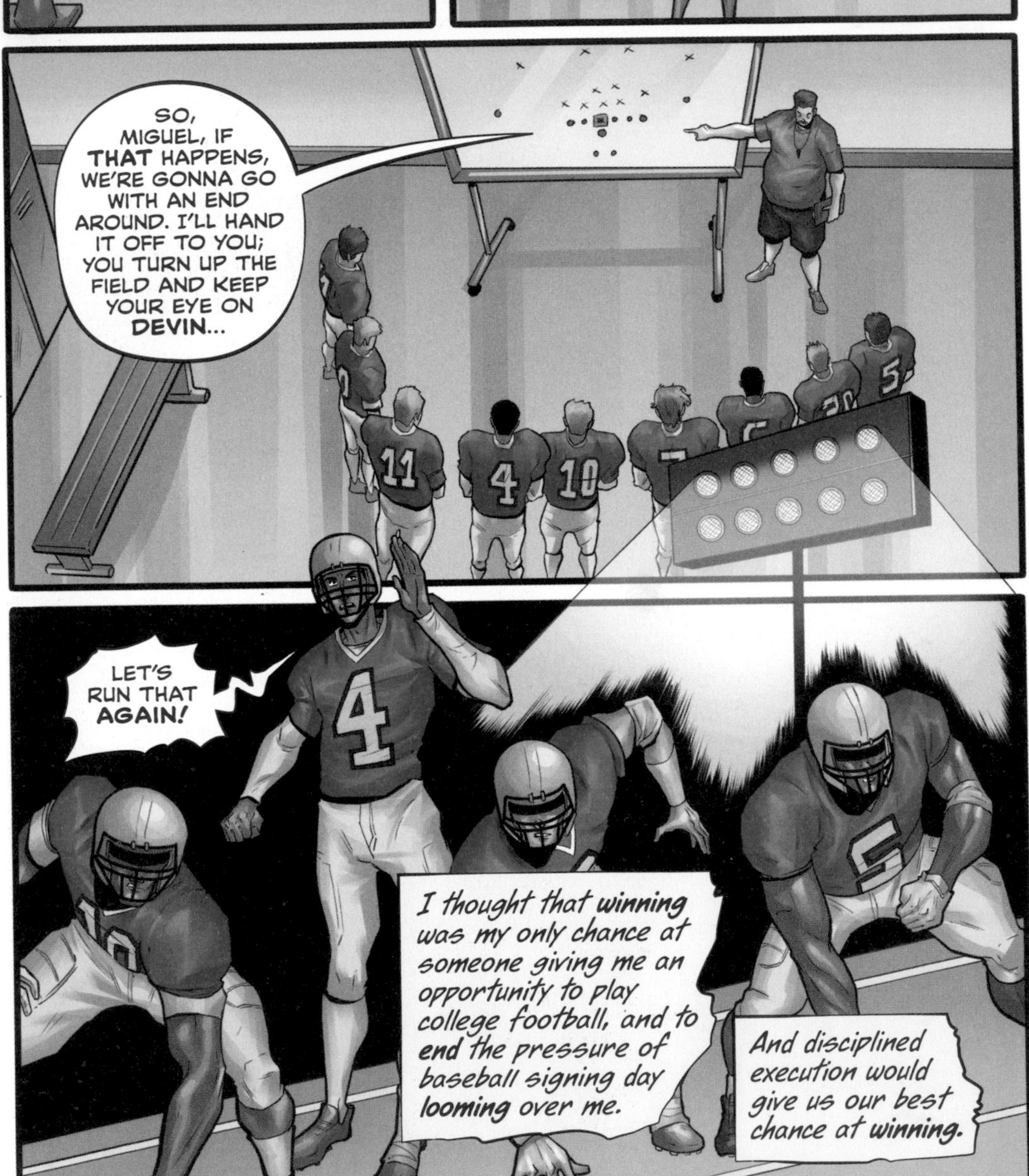
SO, MIGUEL, IF THAT HAPPENS, WE'RE GONNA GO WITH AN END AROUND. I'LL HAND IT OFF TO YOU; YOU TURN UP THE FIELD AND KEEP YOUR EYE ON DEVIN...
LET'S RUN THAT AGAIN!
I thought that winning was my only chance at someone giving me an opportunity to play college football, and to end the pressure of baseball signing day looming over me.
And disciplined execution would give us our best chance at winning.

I had talked a big game because I knew my team needed the confidence.
Truthfully, though, when the game against Lincoln came, I felt like we were the underdogs.
14
10
7
21
5
2
4
But the way life works, when a challenge comes along, you step up.
LET'S DO THIS.

Our defense had been grinding nonstop to prepare for this moment.
And it seemed like it might not be enough.
IT'S TIME TO LOCK IN! LET'S GO!
FOCUS.
...we had to let that go.

AND WITH THAT, THE TROJANS HAVE TAKEN A FOUR-POINT LEAD!
It felt like we were fighting for our lives out there.
But we couldn't give up. The idea that they were so much bigger, so much faster...
TOUCH-DOWN! PITMAN WINS!
AND THE PITMAN PRIDE TAKE THE FIRST ROUND OF SECTION PLAYOFFS! WOW!

SO WHAT DO YOU THINK WAS THE SECRET TO THE TEAM'S SUCCESS?
WE STEPPED UP, WE WERE DISCIPLINED, AND THAT'S HOW WE WERE ABLE TO FINISH STRONG!

SO YOU MUST BE PRETTY CONFIDENT HEADING INTO THE GAME AGAINST TRACY WEST?
ONE GAME AT A TIME.
YEAH, WE ARE! LET'S GO!

I had to say that... but deep down, I knew the section title was ours. Everyone knew Lincoln was the hardest team we would go up against.

Everything else would be a walk in the park.

It turns out Tracy West wasn't going down without a fight. By halftime, we were tied.
AND TOUCHDOWN FOR THE PRIDE AFTER A TWENTY-ONE-YARD PASS FROM KAEPERNICK!
It almost seemed like the harder we pushed...
...the more they wanted it.

PASS INTERFERENCE!
There's this thing in a game, where you feel the energy shift. And it was happening. Everyone was getting flustered.
But then I saw my opening. I could save this. Save us.
This was it. Everything we had worked for.

AND THAT'S IT, FOLKS! THE WOLF PACK WINS IT ALL! 27-23!
We lost.
We lost.

It's over.

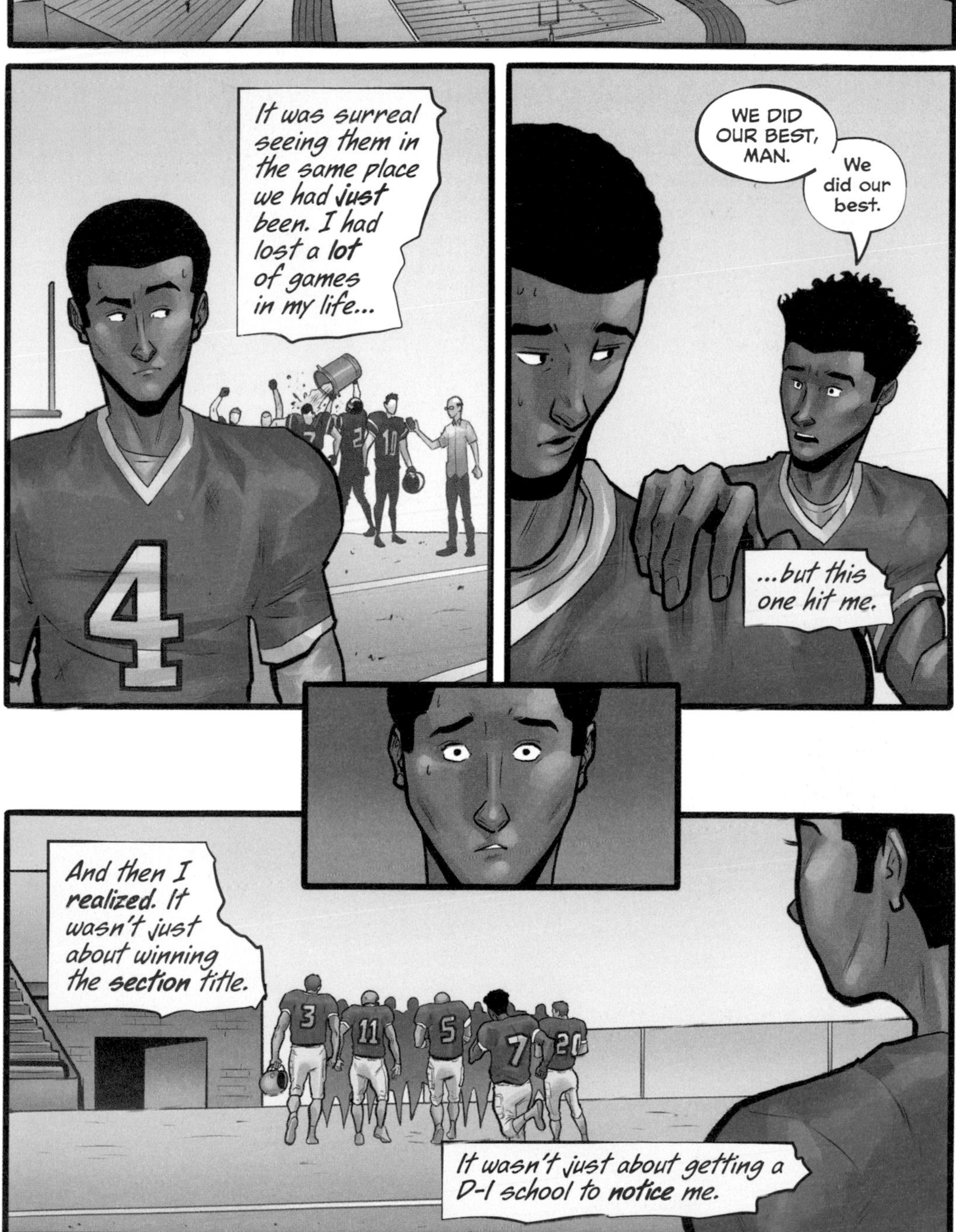
It was surreal seeing them in the same place we had just been. I had lost a lot of games in my life...
WE DID OUR BEST, MAN.
We did our best.
...but this one hit me.
And then I realized. It wasn't just about winning the section title.
It wasn't just about getting a D-I school to notice me.

This game... this was the last game I would ever play with these guys.
The people who made me feel the most secure, the most like myself, of anyone in the world.

My team.

CHAPTER 7
DECISIONS

It was weird...
...Waking up and not having this one goal hanging over my head, driving me forward all day, every day.
I felt awful. But then I had to remind myself...
It was never about that one game.

HEY, THERE'S TURLOCK'S ACE! HOW'S IT GOING, COLIN? TERESA?

Oh, HEY THERE, HANK! WOW, ANGELO IS GETTING SO BIG!
HI, MR. GILPIN–
SO WHAT'S THE WORD, COLIN? YOU SIGNED WITH LSU YET? BOY, THEY COULD USE A PITCHER LIKE YOU.

HE'S STILL FIGURING IT OUT, BUT THANKS FOR YOUR SUPPORT.
GO, TIGERS!

ALL DONE WITH THE QUIZ?
YEP. AND LUCKY ME, BECAUSE I HAVE A **MATH** TEST NEXT PERIOD, TOO, SO I'M GONNA STUDY A LITTLE.

ACTUALLY, COLIN, MAY I SPEAK WITH YOU FOR A MOMENT?

What, do you think he was cheating?
Pffft
You know Colin's got all those Malcolm X speeches memorized. All the marches and stuff.

I SHOULDN'T HEAR ANY TALKING IN THERE!

I SPOKE TO YOUR COUNSELOR, AND SHE TOLD ME YOU HAVEN'T **SIGNED** YET.
NOW, I KNOW THIS IS A TOUGH **DECISION**, AND YOU'RE HOLDING OUT FOR **FOOTBALL**, BUT IT'S IMPORTANT TO REMEMBER THAT THIS IS YOUR **FUTURE** WE'RE TALKING ABOUT. AND COLLEGE IS THE KEY TO...
To everyone else, the writing on the wall seemed **clear**. I had countless baseball prospects and exactly **zero** football offers.
As far as they were concerned, my future was on the **mound**.

I wasn't so sure.

For one thing, before I could get to the major league, I most likely would have to join the farm leagues.
I knew that life could be tough for those guys. Playing constantly, no days off to rest.
It would be a grind to make it... and I knew I could do it...
...but there was something else.
PITMAN
PITMAN
PITMAN

I'm starving. I'm gonna get a doughnut later from O'Mally's.
Oh man, did you hear Mr. O'Mally sold the place? New ownership.
Oh yeah? To who?
PITMAN

Some **Mexican** stealing another job. Same menu, though.
P

I learned a long time ago that even if it's uncomfortable and even if you're standing alone...
...even when the path is unclear...
...you have to be able to stand up for what's right.
PITMAN

HEY.
THAT'S NOT COOL.

WHAT'S NOT COOL?
WHAT YOU SAID IS REALLY MESSED UP.

I WAS JUST JOKING AROUND, MAN. YOU DON'T HAVE TO MAKE A WHOLE BIG DEAL OUT OF IT.
OH YEAH? WELL, I AM. YOU'RE BEING A RACIST PIECE OF–

HEY! KAEPERNICK! WHO TOLD YOU TO STAND UP?
SORRY, COACH, BUT HE SAID–
I DON'T CARE WHAT HE SAID. ARE YOU IN KINDERGARTEN? STICKS AND STONES!

BUT HE–
I SAID I DON'T WANT TO HEAR ABOUT IT!

"Sticks and stones." I couldn't stop thinking about it. I was so angry. If I didn't speak up, it felt like something would eat at me inside. But when I did speak up, apparently I was doing the wrong thing.
I didn't know what to do... except what I always did.
Show up and work it out.
STOP, DROP, SHUT 'EM DOWN, OPEN UP SHOP
Hey!
HEY!
SORRY TO SCARE YOU, EARLY BIRD!
Ah, SORRY! I DIDN'T HEAR YOU.
ALL GOOD. I FIGURED YOU WERE JUST TRYING TO FORCE ME TO HIT YOUR PACE.
LET'S KEEP GOING!

I DIDN'T EXPECT TO SEE ANYBODY ELSE OUT HERE THIS EARLY.
I'M HAPPY TO SEE YOU.
I'M... REALLY HAPPY TO SEE YOU, TOO.
I wanted to say, "Really, really, really happy."
I wanted to say, "I needed this."
But I kept that to myself.

I FEEL LIKE PEOPLE AROUND HERE MAKE IT PRETTY **CLEAR** THAT THIS TOWN IS NOT OUR TOWN. NO MATTER HOW FAST WE RUN, OR HOW **GOOD** WE **BALL**.
NO MATTER HOW **LOUD** THEY CHEER.
BUT YOU KNOW WHAT, COLIN? **THEY** DON'T GET TO **DECIDE**...
WHO WE ARE. OR WHAT WE'RE **WORTH**.
LET'S TAKE A BREAK.
40
30
20
10
YEAH, COLIN. I THINK THAT A LOT.
HEY, DO YOU EVER FEEL LIKE...
You know. I don't know how to **say** this.
KIND OF LIKE TURLOCK WASN'T REALLY **MADE** FOR US?
YOU OKAY?
YEAH, JUST... THINKING. AND CATCHING MY **BREATH**.

BUT... YEAH. THAT'S WHY I **WORK** SO HARD, YOU KNOW?
I FEEL LIKE... IF I'M HONEST WITH MYSELF, I'M **GOOD**. I **KNOW** I'M GOOD.
I FEEL LIKE IF I PUSH THAT, IF I KEEP GOING, I HAVE A REAL CHANCE.
TO SPEND MY LIFE DOING SOMETHING I **LOVE**. SURROUNDED BY PEOPLE WHO... **GET** ME.

DO YOU KNOW WHAT I **MEAN**?
I KNOW **EXACTLY** WHAT YOU MEAN.

MORNING.
MORNING!

HEY, TIFFANY...

WOULD YOU GO TO THE SPRING DANCE WITH ME?

I JUST DON'T SEE WHAT'S WRONG WITH **LINDSEY**!
I NEVER SAID THERE WAS ANYTHING WRONG WITH LINDSEY.
I DON'T KNOW WHY YOU'RE MAKING SUCH A BIG **DEAL** ABOUT THIS.

I JUST THINK SHE'S A **REALLY** NICE GIRL, AND I THINK YOU SHOULD GIVE HER A **CHANCE**.

AND I THINK THAT IT IS **WEIRD** FOR YOUR **MOM** TO DECIDE WHO YOU TAKE TO A HIGH SCHOOL DANCE.
DON'T EAT THOSE—YOU'LL **RUIN** YOUR APPETITE!

LISTEN TO YOUR **MOTHER**, COLIN.

THESE COOKIES **WILL** RUIN YOUR APPETITE.

THAT'S NOT THE **ISSUE**!
MOM THINKS I SHOULD TAKE LINDSEY KELLER TO THE SPRING DANCE.
I ALREADY ASKED **TIFFANY**. BECAUSE THAT'S WHO I WANT TO TAKE.

AND I DON'T UNDERSTAND WHY THIS IS A WHOLE BIG **THING**.
I JUST... I just... LINDSEY COMES FROM SUCH A NICE **FAMILY**.

AND TIFFANY DOESN'T?
THAT'S NOT WHAT I'M **SAYING**.
HEY, WHAT HAPPENED TO **MATH** HOMEWORK?

I CAN'T CONCENTRATE.

Oh, here we go again with the BET.

SO THE GIRL I WANT TO TAKE TO THE DANCE, **THAT'S** A PROBLEM.
WHAT I WANT TO WATCH ON TV, **THAT'S** A PROBLEM.
ANYTHING ELSE?

OH, STOP BEING SO **DRAMATIC**.
WHAT'S GOTTEN **INTO** YOU LATELY?

YOU KNOW WHAT? I'M JUST GONNA TRY TO DO MY HOMEWORK **UPSTAIRS**.

SERIOUSLY, COLIN. WHAT'S GOING ON?
REFUSING TO MAKE A **SIGNING DECISION**, PLAYING AROUND WITH YOUR FUTURE, AND MAYBE MISSING A **BIG** OPPORTUNITY. WE DON'T WANT TO FORCE YOU EITHER WAY, BUT YOU HAVE TO MAKE **SOME** DECISION, AND WE'RE GETTING CLOSE TO THE WIRE HERE.

AND COACH ROGERS CALLING US, COMPLAINING YOU'RE ACTING OUT ON THE **BUS** LIKE A LITTLE KID.
WHAT IS IT, COLIN?

YOU WOULDN'T UNDERSTAND.
NOW, MAY I BE **EXCUSED**?

But–
DAD.

I AM ABOUT TO GRADUATE HIGH SCHOOL.
I WOULD LIKE TO BE ABLE TO WATCH THE MUSIC VIDEOS THAT I WANT TO WATCH.
AND I WILL BE TAKING THE GIRL THAT I LIKE TO THE SPRING DANCE. THAT GIRL IS TIFFANY. NOT LINDSEY.
MATH
AND AS FOR FOOTBALL?
FOOTBALL IS WHERE I SEE MY FUTURE.
AND THAT IS A PATH I HAVE TO CHOOSE FOR MYSELF.
I'M GOING TO GO FINISH MY HOMEWORK NOW.
THANKS.

CHAPTER 8
COMMITMENT

YOU GOT THIS!
The season was over. But I was still holding out hope that my football career wasn't.

YOU SURE YOU'RE GOOD?
YEAH, I'M JUST GONNA STICK AROUND A LITTLE LONGER.
I told myself that I had to stay ready. Stay focused.
That even as baseball signing day was looming, a D-I football team was going to come calling any day now. I tried not to think about what it would mean if they didn't.

Any day now.

Working out with my QB coach became my favorite thing to do. When I was out there, I could **forget** about **everything else.**
Usually, this one guy who played for Cal was there– Damon Matthews.
Man, I looked up to him so much.

To me, Damon represented everything I **wanted** to be in football.
Smart as anything.
The ability to make everybody **laugh** and to talk a good game, but also to **play hard.**

GO!

NICE! NICE!
FOOTWORK LIKE THAT? YOU LOOK COLLEGE-READY.
MAN... I'M GLAD YOU THINK SO.

NO BITES?
NOT A ONE.

SORRY, BRO. BUT A BIT OF ADVICE: YOU CAN ONLY CONTROL WHAT YOU CAN CONTROL, YOU KNOW?
PLAY YOUR BEST GAME. EVERY GAME. ON AND OFF THE FIELD. GET RIGHT WITH YOURSELF, GIVE IT ALL YOU'VE GOT, AND THE REST WILL FOLLOW.

OR IT WON'T. BUT EITHER WAY, YOU GOTTA BE ABLE TO KNOW YOU GAVE IT YOUR ALL.
KAEPERNICK! MATTHEWS! ENOUGH CHITCHAT!
SORRY, COACH!

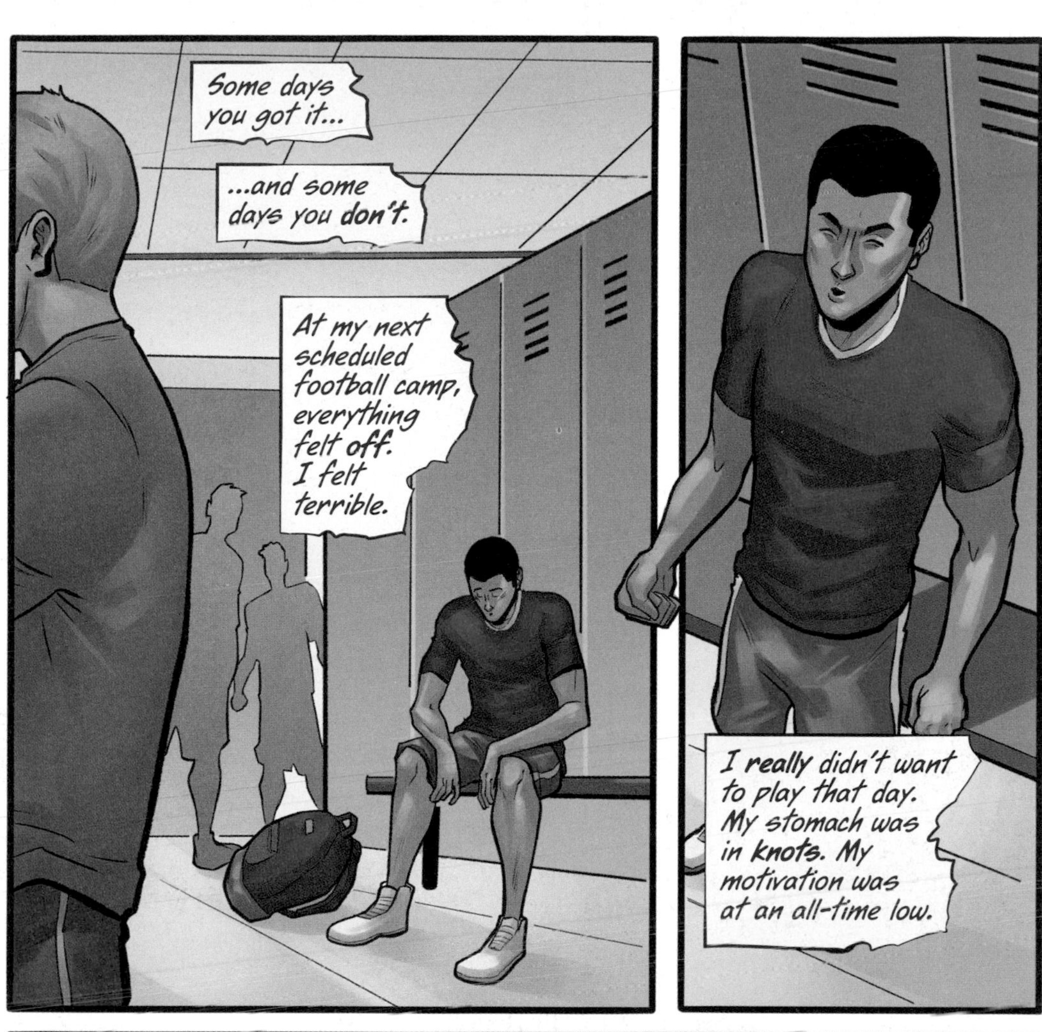
Some days you got it...
...and some days you don't.
At my next scheduled football camp, everything felt off. I felt terrible.
I really didn't want to play that day. My stomach was in knots. My motivation was at an all-time low.

YOU GOOD, COLIN?
I'M GOOD. LET'S HEAD OUT.
I just kept thinking of Damon's words.
"Best game. Every game."

"Get right with yourself."
I felt like I was finally getting there.

NICE WORK, SON.
THANKS, COACH.

KAEPERNICK?

JED WILSON. **UNIVERSITY OF NEVADA, RENO.** YOU GOT FIVE MINUTES?
*By that point, I'd had **hundreds** of scout conversations and still no football offers.*
I figured the best thing was not to think about it.

Besides, I had other things on my mind.

I'M HEADING OUT TO PICK UP TIFFANY FOR THE DANCE!
OKAY, HAVE FUN!

Nice to meet you.
Nice to meet you!
Ahem SO NICE TO MEET YOU.

DING DONG

YOU MUST BE COLIN!
NICE TO–

COME ON IN!
COME ON IN, SON!

SO NICE TO MEET YOU, COLIN!

I'M TIFFANY'S MOM, SANDRA! THIS IS MY HUSBAND, JEFFREY.
OUR SONS, JEFFREY JR.— we call him JJ— AND MICHAEL.
AND THAT'S TIFFANY'S AUNTIE CHERISE. AND YOU ALREADY MET UNCLE RONALD.
IT'S REALLY NICE TO MEET ALL OF YOU!

AND LOOK WHO MAKES HER APPEARANCE!
HI, COLIN. I SEE YOU MET THE FAMILY.

NOW, YOU ALL CAN'T LEAVE UNTIL YOU EAT SOMETHING. THEY NEVER HAVE DECENT FOOD AT THESE THINGS!
THEY'RE TRYING TO GET TO THE DANCE, AUNTIE CHERISE!
THAT CAN WAIT!
TIFFANY, YOU LOOK BEAUTIFUL. LET ME GRAB MY CAMERA!

BYE!
HAVE FUN!

IS IT WEIRD TO SAY THAT I COULD HAVE SPENT THE WHOLE NIGHT HANGING OUT WITH YOU AND YOUR FAMILY?
NAH. THEY'RE PRETTY COOL! THEY LOVED YOU. YOU'RE WELCOME ANY TIME.

COLIN... I DON'T KNOW HOW TO TELL YOU THIS, BUT...
Mhm?

YOU HAVE BARBECUE SAUCE ON YOUR CHIN. AUNTIE CHERISE GOT YOU GOOD, HUH?

Growing up, I constantly had people telling me, "You're a football player" or "You're a baseball player."
I never saw it **that** way.

I **played** baseball. I **played** football. But I never wanted to tie my identity to that.
I wanted to be around people who could see me... for **me**.

And sometimes in my life, there were even people who helped me see parts of myself that I **didn't** see, or didn't fully understand yet.

I'll always be thankful for that.

"FREEING YOURSELF WAS ONE THING...
"CLAIMING OWNERSHIP OF THAT FREED SELF WAS ANOTHER."

WHAT DO WE THINK MORRISON MEANT BY THAT?

COLIN?
I THINK... YOU CAN MAKE A DECLARATION, IN YOUR HEART, ABOUT WHO YOU WANT TO BE.
BUT THEN YOU HAVE TO REFLECT THAT IN YOUR ACTIONS. YOU HAVE TO MAKE IT REAL.

THAT'S GREAT.
BUT THEY WANT YOU DOWN AT THE MAIN OFFICE.

HI, I GOT CALLED DOWN HERE?

YES, YOU HAVE A CALL!
YOU CAN TAKE IT IN MY OFFICE.

COLIN, THIS IS **CHRIS AULT** CALLING YOU FROM THE UNIVERSITY OF NEVADA, RENO. I'M THE HEAD COACH. SON, I'D LIKE TO OFFER YOU A FOOTBALL SCHOLARSHIP.
COLIN? COLIN, ARE YOU THERE?
YES, I'M HERE. I'D LOVE TO COME PLAY FOR YOU, COACH.

That day, I knew I was starting the next chapter of my life.

That day, I saw someone new in the mirror. Someone **growing**. A D-I football player. A college student.
Someone with the clarity and **self-knowledge** to walk his own **path**, his own **way**, no matter how uncertain.

Even now, I don't pretend to know everything.
But I know some things for sure.

COLIN KAEPERNICK's Know Your Rights Camps' mission is to advance the liberation and well-being of Black and Brown communities through education, self-empowerment, mass-mobilization, and the creation of new systems that elevate the next generation of change leaders.

On April 23, 2022, Colin hosted a Know Your Rights Camp for youth in Las Vegas, Nevada. He asked campers to share how they aspire to "change the game." Here's their vision for a better future.

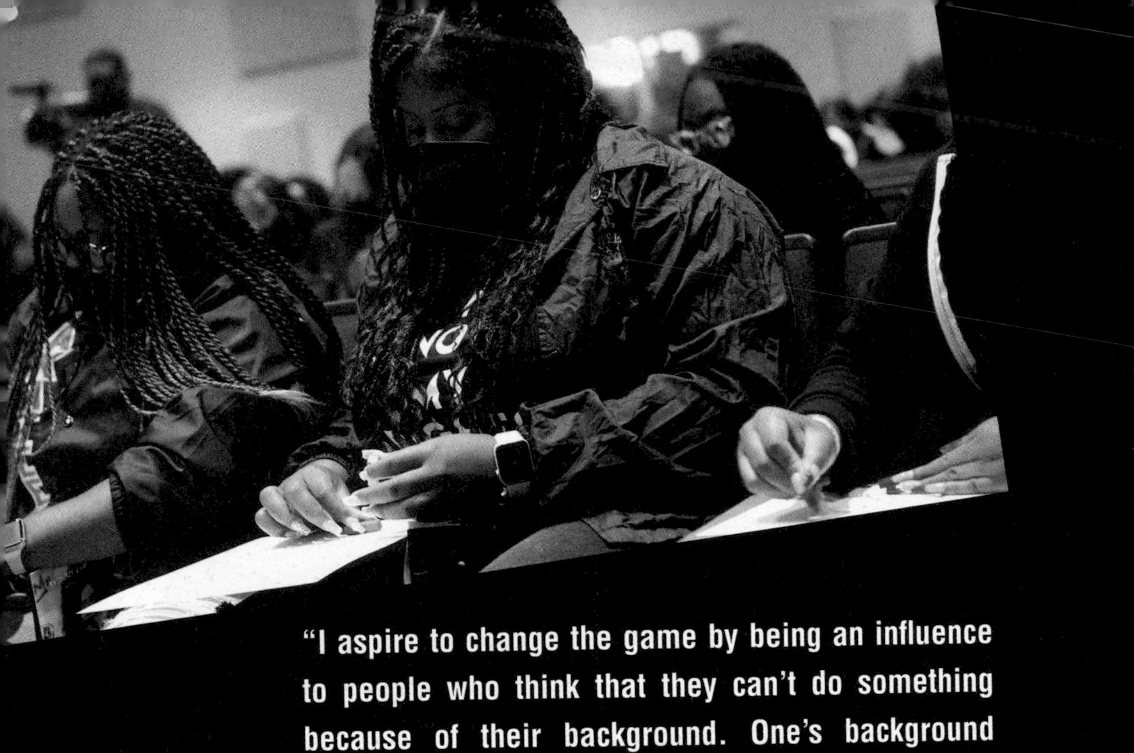

"I aspire to change the game by being an influence to people who think that they can't do something because of their background. One's background shouldn't matter at all and you should strive for what you want no matter what. Be your own person. Don't follow the leader—be the leader!" —MYA DAVIS

"I aspire to change the game by becoming an influential Black broadcast journalist."
—CAMERON COBBIN

I KNOW MY RIGHTS

“I aspire to change the game by furthering my knowledge to become a future educator to youth in schools.” —BREANNA ONORI

“I aspire to change the game by doing more for my community and for female athletes.” —ZHANINA BURRELL

"I aspire to change the game by using any and every platform I can to educate people of color so I can pour back into the community that poured so much into me." —LAURYN BURIST

"I aspire to change the game by investing in my community and helping to build a community of leaders driven to change the world." —KATIE MEMBRENO

KNOW YOUR RIGHTS
10 POINTS
1. YOU HAVE THE RIGHT TO BE FREE.
2. YOU HAVE THE RIGHT TO BE HEALTHY.
3. YOU HAVE THE RIGHT TO BE BRILLIANT.
4. YOU HAVE THE RIGHT TO BE SAFE.
5. YOU HAVE THE RIGHT TO BE LOVED.
KNOW YOUR
RGHTS CAMP
10 POINTS
HAVE THE RIGHT TO BE FREE.
HAVE THE RIGHT TO BE HEALTHY.
HAVE THE RIGHT TO BE BRILLIANT.
OU HAVE THE RIGHT TO BE SAFE.
OU HAVE THE RIGHT TO BE LOVED.
OU HAVE THE RIGHT TO BE ALIVE.
I
KNOW
KYRC STAFF

EVE L. EWING is a writer and scholar from Chicago. She is the award-winning author of the poetry collections *Electric Arches* and *1919*; the nonfiction work *Ghosts in the Schoolyard: Racism and School Closings on Chicago's South Side*; and a novel for young readers, *Maya and the Robot*. She is the co-author (with Nate Marshall) of the play *No Blue Memories: The Life of Gwendolyn Brooks*. She has written several projects for Marvel Comics, most notably the *Ironheart* series. She also cowrote the short story "Timebox" with Janelle Monáe for the collection *The Memory Librarian*. Her work has also been published in the *New Yorker*, the *Atlantic*, the *New York Times*, and many other places. She is a professor at the University of Chicago.

Credit: Valerie Zamora

Colombian artist **ORLANDO CAICEDO** grew up eating arepas and drawing. After completing his BFA with a concentration in illustration at the Atlanta College of Art, he decided to do that full-time. He's the winner of Stan Lee's POW! Entertainment/LINE Webtoon Superhero Comics Contest as well as the Mad Cave Talent Search. His published work includes Webtoon's The *Badguys* and *Pound*, and Mad Cave Studios' *Dry Foot*. He worked In animation before shifting to comics on hit shows such as the Emmy Award–winning *Archer* for FX and *Frisky Dingo* for Adult Swim. When not drawing comics, he can be found napping with his loving wife and new baby daughter. You can visit him online at www.ocaicedo.com.

COLIN KAEPERNICK is a Super Bowl quarterback and *New York Times* and *USA Today* bestselling author who fights oppression globally. He founded the Know Your Rights Camp, which advances the liberation and well-being of Black and Brown people through education, self-empowerment, mass mobilization, and the creation of new systems that elevate the next generation of change leaders. In 2019, he started Kaepernick Publishing to empower a new generation of writers and creators through the development and publication of meaningful works of all genres that focus on amplifying diverse views and voices.

KaepernickPublishing.com